D0181023

Beauty Fades,
Dumb Is Forever

ALSO BY JUDGE JUDY SHEINDLIN

Don't Pee on My Leg and Tell Me It's Raining

Beauty Fades, Dumb Is Forever

The Making of a Happy Woman

JUDGE JUDY SHEINDLIN

Cliff Street Books
An Imprint of HarperCollinsPublishers

First HarperPerennial edition published 2000.

Designed by Kris Tobiassen

The Library of Congress has catalogued the hardcover edition as follows:

Sheindlin, Judy, 1942–
    Beauty fades, dumb is forever : the making of a happy woman / Judy Sheindlin.
        p. cm.
    ISBN 0-06-019270-4
    1. Women—Psychology. 2. Women—Conduct of life. 3. Self-esteem in women. 4. Self-confidence. 5. Self-reliance. 6. Autonomy (Psychology) I. Title
HQ1206.S47 1999
158.1'082—dc21   98-51364

ISBN 0-06-092991-X (pbk.)

00 01 02 03 04 v/RRD 10 9 8 7 6 5 4 3 2 1

To Jerry, who still makes my heart sing,
and to our daughters . . .
all of our daughters

Contents

Acknowledgments

A big hug to Catherine Whitney, whose genius and skill helped organize my musings.

To Jane Dystel, a dynamite literary agent, who was there for me from the beginning, with her insight and encouragement.

A well-deserved thank-you to Diane Reverand, a superbly talented editor.

To all the women who shared with me the very private corners of their lives. Many thanks for your honesty.

Quite a group of women!

Beauty Fades, Dumb Is Forever

Introduction

My father always told me I was beautiful, even during those periods when I was physically awkward—a little too fat, pimples marching across my face, hair that wouldn't curl. If I had a pimple in the middle of my forehead, he'd say, "Oh, that's a beauty! Everyone should have one." My father was a kidder, and he could usually make me laugh. In laughing, I was able to feel better about myself.

He also gave me a piece of advice when he caught me worrying too much about my looks. "Beauty fades," my father would tell me, "but dumb? Dumb is forever." He impressed upon me that a bright intellect, a curious mind, and a passion for learning were priceless commodities. For that reason I always thought that the worst insult in the world was to be called stupid. Since even the beautiful people, if they were dumb, didn't seem all that attractive, I strove to be smart.

To this day, when I'm on the bench it's second nature for me to point a finger at the side of my head and ask, "Does it say stupid here?"—the implied answer being a resounding no. (This habit is a cleaned-up version of my father's favorite line—"Do you see *schmuck* written here?")

As the years went by and I became a lawyer and then a family court judge, my father's words stayed with me. Day after day in my courtroom, I saw what happened to women who didn't use their heads. Somewhere along the line, these women had decided, maybe subconsciously, to hide their light under the proverbial bushel in order to be more attractive to men. And the fallout of their stupid decision was the daily parade of misery that marched through family court.

In most jurisdictions, ignorance of the law is no excuse, and for these women, ignorance of life was no excuse in my courtroom—and it still isn't.

I've heard it all before: *"I didn't know . . ." "I didn't think . . ." "I didn't mean to . . ."*

That's dumb-talk.

Most of the problems that ended up in family court could somehow be traced back to my father's adage. There were plenty of women only too happy to prove him correct. The price of putting their brains in a vault was very high—not just for them, but for their children and their children's children.

I realized that if things were going to change, we had to get the message out to girls, early on, before they started making irreversible choices, that beauty fades . . . but dumb? Dumb is forever.

LET'S TALK

I don't presume to be an expert on marriage, parenting, or all the issues women face. I have, however, dealt with thousands of women representing all age groups and circumstances during my years serving as a judge in family court. I think I am something of an expert on why women make stupid choices.

I've also made a few beauts of my own, which, in the interest of literary accuracy, I feel compelled to share with you.

I've never hesitated to say what I think, to talk straight. I'm basically a positive person. I am not a victim. That confidence is a gift I was given by my parents, and it's something I want to pass on to other women.

I don't take Saint John's Wort or Prozac. If I have a bad day, then I have a bad day—everybody has a bad day once in a while. I accept it. I'm grateful that my bad days are less frequent than my good days. If things aren't going my way, I don't pull the covers over my head and stay in bed. Life goes on. I get up, take a shower, and clean the kitchen. That's how I was raised. "You'll feel better. Get on with it."

I understand that life has those moments when you think, "I'm going to die of embarrassment"—or shame, or regret. Times when everything just seems too hard. But I've lived long enough to know that those times pass, and this knowledge gives me a positive attitude. I'm always looking for the surprise, the new opportunity, the adventure. I have only one life to live—and I'm not going to waste it. I'd rather burn out than rust out.

I have some insight, because I've seen what failure can do. I've seen what the lack of self-worth does to people, especially to women. Women put themselves in bad business situations, and they tolerate demoralizing and demeaning conditions. They don't choose their mates; they allow themselves to be chosen. They stay in relationships that are untenable, with men who are physically and emotionally abusive. They defer to men, because they doubt their own worth.

Self-worth is directly related to confidence—that "can-do" attitude. When you feel as if you're in control of your life, you have a sense of security that pervades everything—from the profound to the mundane.

In the category of mundane, I'll give you a simple example. For years, when the needle of my car's gas tank neared empty, I pulled into a gas station and said to the attendant, "Please fill it up with premium—thank you." He'd top off the tank, kick the tires, check the oil, wash the windows, and I'd be on my way.

Then came the world of self-serve. After a lifetime of service with a smile, they now expected me to pump my own gas. Initially, I resisted. It wasn't that I was a princess, afraid of getting my hands dirty—although I didn't relish the thought of wearing *eau de premium* as my scent. My problem was that I was intimidated by the whole process. That gas pump was like a high-tech alien.

Which buttons did you push? How did you stop it? What was the secret of hauling the snakelike pump from its pedestal and affixing it to the little hole in the side of the vehicle? I always felt there was a distinct possibility that more gasoline would end up on me than in my car's gas tank. So I would drive for miles out of my way to find a full-service station. And when that didn't work, I would arrange to get gas only when I was with my husband, Jerry.

One day the inevitable happened. It was just me, the car, and the self-serve gasoline pump. And I did it. The freedom, the power, the control that I felt.

I was free from any male tyranny. I could pump my own gas!

Pump once and you can pump forever. Change one tire and you'll always be free of the "fear of the flat."

That's the mundane.

The profound requires more work, but the principle is the same. I'm convinced that independence is a woman's only path to happiness. That doesn't mean you necessarily have to be on your own; the point is knowing that you *could* make it on your own. And the only way to possess this confidence and control is to have a profession or a vocation that gives you pleasure and makes you self-supporting.

When my first marriage ended after twelve years, I was plagued by most of the usual fears that a woman alone with two young children feels. However, the one fear I didn't have was that I wouldn't be able to support myself and my children.

I wasn't sure I'd ever find someone else to share my life with. I wasn't sure I could handle the car in a really bad snowstorm. I was absolutely sure that, if I needed a washer changed in the sink, I was helpless.

But I always knew that no matter what else happened, I'd be able to earn a living doing something that I really enjoyed. Confidence equals self-worth, and self-worth equals control. I had control of my destiny where it really mattered. And that, in a nutshell, is the definition of self-esteem.

JUDGE JUDY'S LESSONS

Today I'm a fifty-five-year-old woman. I've been a lawyer for thirty-four years. I've been a daughter, a wife, a mother—and now I'm a television personality.

I've seen a lot!

As a lawyer and then a judge in the family court, I represented and later adjudicated cases involving tens of thousands of women. All these women were in a state of crisis:

- The young and pregnant.
- The divorced and penniless.
- The physically abused.
- The emotionally abused.
- The poor and homeless.
- The addicted and defeated.
- The enablers of abuse.
- The deliberately blind—who looked away while fathers or boyfriends abused their daughters sexually.

There was one common denominator—a lack of self-worth. Nobody had ever told these women, in a way they could believe, that they were important in their own right. We in the courts were left to throw lifelines to women who had already drowned and were taking their kids down with them.

I was fortunate. I enjoyed the gift of two parents who were committed to providing me with the building blocks to a secure future. My childhood was memorialized in endless photographs, and I knew that they loved me. They imbued me with an ethical and moral code of the highest values, and actively participated in my life. They continually reinforced the belief that I could do anything I set my mind to—and I believed them.

Being a wife and mother was a bit trickier, because, like everyone else, I was on my own by then. I freely admit that I made a lot of mistakes. Some were trivial, others not so trivial, but the important thing was that I learned from my mistakes. They all served as valuable lessons.

In the following pages, I am going to impart the simple and true lessons I've learned, both on and off the bench. They get to the heart of the matter for all women, no matter what your place in life. They are:

1: **Beauty fades, dumb is forever.** Don't allow yourself to be defined by the way you look. Remember, a gorgeous model will someday become a "former model." A beauty queen will eventually be an "aging beauty queen." But if you've got brains, spirit, drive, and personality, you'll never be a "former" anything.

You'll always be the one great you. Don't hitch your identity and your dreams to a facade. Build a solid foundation—from within.

2: **Don't crawl when you can fly.** If you have talent, let it show. If you have skills, go ahead and excel. Stop worrying that you might come on too strong, be too aggressive, be not feminine enough. I'm talking about that mask we women wear in the presence of men. From the bedroom to the boardroom, we defer to men. It's our dirty little secret. I say it's time we dust off our knees!

3: **What goes up must come down.** Sex does *not* give women power over men. So why do so many behave as if it does?

We think if we can give a man what he wants in the bedroom, all the rest will follow. I'm assuming that men invented this lie, since it serves their purposes so well and ours so poorly. Remember, ladies, your body is your temple. It belongs to you. If you let someone inside, it should be for the right reason— not for a Saturday night date or to avoid a fight.

4: **Denial is a river in Egypt.** Get it? Women are master deniers. They're the only known species who can be covered with bruises and still think, "But he really loves me." If your husband, your boyfriend, your boss, your coworker, or your friend is abusing you, you're a victim. But you're also a dope.

It's time for an eye-opener.

5: **Master the game—then *play* it.** If you want to succeed in the workplace, learn the rules and play the game better than men. The greatest obstacle to success for women is that while the men are busy looking out for themselves, most of the women are busy looking out for the men, too.

6: **You're the trunk of the tree.** In every human endeavor, be it family, community, work, church, or school, you will find women holding everything together. We are the trunk—the foundation, the source. But a tree needs nourishment to stay strong, and you're not usually going to get it from the outside.

Start carrying your own pitcher of self-esteem so you can nourish yourself.

7: **You can't teach the bull to dance.** Don't spend your precious life trying to force your mate to change. He can't; he won't. You'll only increase your own level of irritation. Make a mental short list of what's important. Fight for those items, and don't machinate over the minutiae.

8: **Failure doesn't build character.** Whoever came up with that idea was an idiot. Success builds character. If we want our daughters to have self-esteem, we need to give them the opportunity to shine. Make them believe they are special, and they *will* be special.

9: **Letting go is half the fun.** It's wonderful and fulfilling to have children—to watch them grow and learn. But parenting is something like spinning a top. If you wind it up to full speed and keep hanging on, it will never spin freely. Don't live your life through your children. If you want them to prosper, you have to let them go—at full speed. And if *you* want to be content, you have to live your life for *you*.

10: **You can be the hero of your own story.** At every age, you have the opportunity to star in your own adventure. Being fifty can be fabulous; sixty sensational. Every age offers something you didn't have before—and I'm *not* talking about lines and liver spots. The story of your life is about one thing only—*you*. What you make of it—alone or with a partner, with children or childless, pretty or plain, wealthy or middle-class—will be the tale that's written. It's your choice how the story ends.

This book is called *Beauty Fades, Dumb Is Forever: The Making of a Happy Woman*. I believe that if you learn these ten lessons, you *will* be happy, no matter what hand you're dealt on the beauty scale.

I don't discount that good looks are a plus. Billion-dollar industries flourish on extolling the perks of good looks. It seems to me that doing the most with what you physically have is important for your self-esteem, and therefore builds confidence. There are very few Elizabeth Taylors out there—and even Elizabeth Taylor isn't thirty anymore. We've got to rely on our brains, our wits, our skills. We need to concentrate on making wise career decisions and smart choices in our mates. Maintaining self-esteem and a sense of humor, and keeping life's inevitable challenges and frustrations in perspective, helps a lot, too. It's a fabulous world out there. Half the world's population are women. It's time for us to have at least half the fun!

CHAPTER 1

Don't Crawl
When You Can Fly

When I was ten years old, my family moved from an apartment into a private house, which was a pretty big deal in those days. The house was about a twenty-minute bus ride from the old neighborhood in which I had grown up, and where all my friends were still living. I'd left behind not only friends, but most of my extended family—my grandparents, my aunts and uncles, and my cousin Shirley, with whom I'd always been close.

Shirley and I were the same age, but you'd never have known it. She had developed early, and was sprouting breasts. I was as flat as a pancake. One Saturday, Shirley invited me back to the old neighborhood for a visit. I took the bus, and she met me at the bus stop. We walked to the high school playing field to meet some of Shirley's friends. On the way, she explained to me that if you wanted to be popular, if you wanted to have a boyfriend, you had to let a boy touch your breasts.

I had never done anything like this before—in part because I had nothing to touch. I was scared to death, but I let a boy touch my breasts that day—such as they were. And that was my introduction, at a very young age, to the way women defer to men.

This is the dirty little secret that women share and rarely talk about. You can be the president of a corporation, an astronaut, a neurosurgeon, a judge—it doesn't matter. All of us started from the same ground zero with the lesson:

If you want to get along in life, you'd better defer to men.

And in different ways, with different levels of success, we all spend our lives trying to push that boulder up the hill. I think there is a better way. In the workplace, I've seen women shrink in silence as a male coworker shared a brilliant idea that, moments before, she had told him in confidence. Does she stand up and say, "That's my idea!" God forbid. How would she be repaid? Perhaps she'd be castigated for not being a "team player," called a "bitch," or even worse, people would stop *liking* her. So, rather than defending her intellectual property, she defers, and lets her colleague take the credit.

There was a time in my professional life when I found myself in that very position. I had a lot of ideas and I was happy to share. My male colleague was just as happy to take the credit. I began to feel like a fool—especially when my colleague got a promotion and I didn't. I decided that I could run the risk of not being liked, but I couldn't stand living trapped in the body of an idiot. So I spoke up, and I'm still here. By the way, it felt terrific!

On the domestic front it was much the same. I did unpopular household chores rather than opening my mouth and creating friction. Very few people love doing the laundry, changing sheets, returning social phone calls, all of the little things that make up a life. I always handled them because it was expedient, my mate didn't want to be bothered, and it was a hell of a lot easier than dealing with a sourpuss. Then I stopped, demanded that we share the load, endured the expected grousing and pouting, and in the end—we're still together.

If you spend your life deferring to someone else, you lose yourself in service. Is that what you want out of your life? It's a high price to pay just to be liked.

Man Craziness

On my television program, *Judge Judy*, all the cases are real, and the parties agree to abide by my decisions. This is life in the nitty-gritty mundane. So many of the cases are about the petty, greedy, selfish, stupid things people do to one another. A large percentage of the cases are demonstrations of women who make outlandish choices for the sake of having a man. Cases in point:

A woman appears before me in court. Her new boyfriend ran up a $3,000 bill on her credit card. I ask, "How long had you been seeing each other before you gave him your credit card?"

"Two minutes and thirteen seconds, your honor."

Well, maybe I exaggerate, but what does it matter? A day, a week, a month—who in her right mind would hand over her credit card to a virtual stranger? She trusted him, she claims. That's a lie. She never trusted him, not deep down. Betrayal is the story of her life. Thanks to her choice of boyfriends, she's always in debt.

Then there's the gal who comes before me who has let a guy move into her house after knowing him for a week. She is a successful businesswoman with a good head on her shoulders—except for this. She describes how every night her new housemate would get duded up, borrow her car, and go out—*on business!* She was supportive and patient. She was understanding. He was trying to make it in the restaurant business. That's why all his meetings took place after midnight. And she gave him the keys, her gas card, whatever. I wasn't born yesterday. All I had to do was take one look at this guy and I could read him like a bad novel. He wasn't and never had been the least bit interested in this very good-looking, middle-aged woman, who moved him into her home because she thought they had a relationship and perhaps a future together. He had the audacity to stand in front of me and proclaim that he'd made it clear to her from the start that they were just friends. She supports him, lends him money to start his own business, gives him her fancy car to drive every night, because she is a charitable woman. He is the aggrieved party. "She wanted more. She knew the deal. She's trying to get back at me because I don't want her."

This guy was a user, and I was frustrated with the woman. Every person in the courtroom, the millions of viewers, all knew exactly what this was about—except for her. She'd decided to play dumb. In these situations, the women always say the same things—"I thought he loved me. He said that he loved me. He promised we'd get married." The check is in the mail. It was the same scenario back in family court. All variations on a recurring theme— *Why did you let him beat you?* "*I thought he loved me.*"

Why did you have his baby? "*He said he loved me.*"

The need to be a couple, to have a mate, regardless of the cost, is the engine that drives the train. At least, that's true for women. It doesn't seem to be the case for men. Would any man suffer regular bumps and bruises to stay in a relationship? Not a chance.

The reason for this puzzling dichotomy is simple. Women are taught that unless they are half of a couple, they are nothing. If you don't believe it, think about the different ways single men and women are described.

He's a confirmed bachelor.

She's a spinster or an old maid.

He's playing the field.

She can't keep a man.

He's choosy.

She's not chosen.

As long as women allow the fiction that being without a man amounts to exile, and as long as couples reinforce the idea by regarding their single women friends with pity, it's going to stay that way. And women are going to keep on deferring.

THE MASK OF SUBMISSION

Deferring means this: subjugating who you are, what you need, and what you want to find, catch, and hold on to a man—any man. Deep down, many women believe that they'll never attract a man if they show their true selves. To that end, they'll train, work, and study to attain an advanced degree in the sublimation of their natural tendencies to snare a really exceptional mate.

Books and magazine articles constantly crank out an old and familiar message to women—do anything you have to do. Two very sharp young women wrote a smash-hit book called *The Rules,* which is nothing more than a handbook of deference. In other words, it's a compilation of rules to help a woman appear to be the person a man wants her to be. I suppose the authors would argue that it's not about deferring at all; it's about being in control. But isn't manipulating a man to the altar ultimately about positioning yourself to defer to what *he* wants?

Sure—in a dating situation both parties try to be at their best, but as a relationship progresses, they should let their hair down and be themselves. Guys do this naturally, even brag about it—"What you see is what you get, baby." Why should the woman wear a mask? A bad day is a bad day, regardless of gender.

My advice is: Don't play games. Be yourself. Be honest, and if he's not interested, then he's the wrong man for you. You set your standards—a good character, a sense of humor, responsible, sensitive, intelligent, caring—and then let a man come along who has those qualities and treats you with respect.

Ladies, choosing the right mate is going to affect every moment of every day of the rest of your life. Women don't sublimate their frustration with a game of golf or a tennis match. So *you* do the choosing, according to *your* standards.

SWAN SONGS

I knew from a very early age that I didn't want to go through life alone. I wanted to be part of a couple, like a swan. Swans mate for life. They glide along the lake in pairs, their elegant necks craning in perfect sync. Even when I was a young girl, I felt certain that it was natural to be paired. I used to joke that the only way I was going to get out of my parents' house was in a pine box or a wedding dress. I opted for the white dress. But things are not always as they seem, even for swans.

I was exposed to a whole other side of the swan story recently while vacationing with my husband at Kona Village, in Hawaii. I noticed a magnificent swan swimming alone, and commented to the owner of the resort that the swan was gorgeous but looked unhappy and dejected. She told me an interesting story.

"There were two pairs of swans for many years," she said. "Then one of the males was killed by a large dog while it was defending its nest. His female partner was distraught. After a few months, her formerly docile behavior began to change, and she became very aggressive."

I was fascinated.

"She was jealous," the owner explained. "She no longer had a mate, and she couldn't stand watching the other pair. So she killed one of the swans."

"The female," I surmised.

The owner looked at me with a raised eyebrow. "No. She killed the male. She didn't want him for herself. She just couldn't bear the fact that the other female swan still had a mate. So she killed *him*." The owner shook her head sadly. "The other female left once her mate was dead. So this territory is all hers. She's *really* alone now. Isn't that something?"

You think this is a bizarre story? Imagine twenty women shipwrecked on a deserted island. They'd probably adjust and establish a well-ordered community. Now add a man to the mix. Instant disaster. The guy could have two heads and a withered personality, and he'd still be a "hot property." The women could kill each other off until only one remained to mate with the male. Or they could just kill him and retain peace in their community. The swan story made me think about the social pressure most women feel to be a part of a couple. Imagine living your life without envy, regret, unfulfilled longing, and disappointment. Imagine choosing your way of life because it will make you happy.

LEARNING THE HARD WAY

My life has unfolded as though it were an endless series of intersecting highways, leading me from my beginnings to where I am today. From the time I was a little girl, I knew I wanted to be a lawyer. I knew I wanted to be married and have children. By the time I was finished with law school and became a lawyer, I was married.

My first job was as in-house counsel to Cosmair, Inc., the parent company of L'Oreal beauty products, L'Oreal of Paris. They had just started distributing their products on the retail market. I was allegedly hired to deal with product liability cases, along with a young man who'd also recently graduated from law school. My first day on the job, I was given a list of pharmacies to call. I was supposed to solicit orders from them—from lipsticks to hair dyes. I thought they were kidding. I even joked, "No, I'm not *that* kind of solicitor!" But they were serious. They didn't give a telemarketing list to my male cocounsel. Needless to say, I quickly developed a deep aversion to my new position.

So what do you do if you hate your job? Have a baby, of course, and eighteen months later, have another. It turned out that my first husband, Ron, had a rather primitive view of family responsibility. When our kids were still small, he announced that he wouldn't watch them until they could say, "I want a hamburger and fries."

When I suggested that he'd had something to do with bringing the children into the world, he acknowledged his contribution, but said, "It's not in my contract to feed them, to change diapers. When they're weaned—you know, toilet-trained and can talk—maybe then."

That's what he said. I didn't argue with him. I was too flabbergasted. Plus I was exhausted. I had two little children to look after, and no help from my partner. Great. I stayed home and took care of the children. On the weekends, Ron's friends would come by and pick him up to play handball. In the winter, it was football. I resented the hell out of it. I spent the entire week alone with the children, and come the weekend, Ron was out having play dates with his buddies. Not a care in the world.

One Saturday morning, his friends called me an hour after they'd picked him up. There had been an accident during the game, and Ron had broken his leg. He was at the local hospital emergency room getting patched up. They brought him home a couple of hours later with a cast up to his hip and helped him upstairs to our bedroom, so he could rest comfortably. At that time we lived in a small two-story house in Brooklyn, New York, with two little bedrooms and a bathroom upstairs, the kitchen and living room downstairs. About an hour after his friends had put him in his bed, I heard Ron calling for me from upstairs.

"Judy? Judy? Honey?" I heard him cry plaintively.

"What is it, Ron? Are you okay?" I shouted up the stairs sweetly.

"I have to go to the bathroom, honey. Could you give me a hand here?" he implored. "I need help getting up."

"That's not in my contract," I yelled up the stairs cheerfully. "You want to go to the bathroom, crawl."

I tried to use that moment to make my point. It was lost as soon as the words were out of my mouth, because I then ran upstairs and lugged him to the bathroom.

Shortly after this incident, we moved to the suburbs, where I became even more isolated. After four years of crabgrass and *Sesame Street*, I felt as if my brain had atrophied from lack of use. I had an insatiable need to use my legal training and to focus on issues of some importance. I needed to get back into the fray. When the children were still in nursery school, I attended a Bar Association luncheon, where I ran into a former classmate from law school. We started talking, and it turned out that he was running the Manhattan office of the family court. He asked me, "What've you been doing?" and I told him. "Changing a lot of diapers, raising a couple of wonderful kids, going berserk, the usual. What about you?"

He said, "That's perfect. You're exactly who I need in my office. You want a job?"

What could I say? "Yes, thanks, I do. Great! When do I start?"

I returned home elated. When Ron came home for dinner later that evening, I told him, "I'm going back to work in two weeks!"

He almost choked on his food. "What do you mean, you're going back to work in two weeks? What about the kids?"

"I called around this afternoon," I replied, "and I'm going to interview a few daytime caretakers. They can do some of the housework, too."

Ron stared at me, stunned.

Let me tell you something about my first husband. He was a very nice man, but he didn't take my returning to work seriously. He thought it was a phase I was going through. He certainly didn't give my career equal weight with his own, and he made that crystal clear when my schedule started getting tight. When I knew I was going to be late, I asked him to try to make it home early for the children. I'll never forget what he said to me when he felt I'd asked one too many times.

"Judy, you work because you want to, okay? You're playing. But I *have* to work."

That was almost thirty years ago, and I still remember his words. Women have memories like elephants—they never forget.

My husband's attitude was a major threat to my happiness and fulfillment. Finally, when I realized we weren't ever going to be headed in the same direction, I decided to divorce. My parents were very upset, and my father, who'd always been my biggest supporter, shocked me when he pronounced, "Two innocent children shouldn't suffer for what two guilty adults did."

My father wanted me to stay in the marriage even though I was unhappy?

That surprised me, but no one had ever been divorced in our family. To my father, I guess, divorce and failure were synonymous. In a way he was right, although I wished he was on my side, as he'd always been before. My mother's reaction was far more sympathetic. She asked my reasons for wishing to divorce. I told her I wasn't happy. She asked if I'd tried to make the relationship work, and I told her that I had.

"I respect your decision," my mother said. Looking back, I realize that my mother knew how much more a woman's happiness depends on a good marriage. For my father, marriage was only one part of his life. He also had his dentistry practice, his golf games, his evenings with Johnny Carson—a multitude of things that gave him satisfaction and pleasure. For my mother, marriage was her main occupation, and her family was her primary source of satisfaction and pleasure.

In the end, I did as my father asked. To placate him I stayed for another year in a marriage that was already over. It was too late to save the marriage, and so, with some regret, I became the first person in my family to wear the scarlet "D."

As for our children, I already knew from my time in family court that children are always better off spending separate time with two happy, calm parents than they are living in a combat zone. I became a single working parent.

Then I met Jerry Sheindlin.

JERRY

I was working as a prosecutor in family court in Manhattan, at 60 Lafayette Street. That's downtown at the edge of the island. Jerry was a criminal defense lawyer at the time. I'd never run into him, though, since he didn't work on family court cases.

It was a Friday night, and Jerry had just finished a big trial in which he'd been victorious. As I was wrapping up my work for the week, my boss came by my office, stuck his head in the door, and invited me to join the rest of the lawyers for a drink at Peggy Doyle's, a local watering hole favored by the legal crowd.

I usually left the office by 5:30 P.M. for the trek to Hartsdale. But this was a Friday. My parents had made it their habit to come up every Friday to spend the day and night visiting. My father's dentistry practice allowed him Fridays off, so he liked to come up to golf. It gave my mother a chance to fuss over her grandchildren. If I wanted to stay out a little late after work on a Friday, I could. Nevertheless, I declined my boss's invitation. It had been a long week, and I really did want just to go home, put my feet up, and be with my parents and my children. But when I left the building, I decided on a whim to stop by Peggy's for a few minutes.

I walked into the bar, and there, sitting at the center of the group, was Jerry Sheindlin, regaling everyone with the colorful tale of his victory in court earlier that day. All my friends were standing around him, enjoying his blow-by-blow description of the case he'd won. Jerry was a great storyteller; he still is.

I took one look at this guy, and something moved inside me. I wanted to meet him. So, in my typical delicate manner, I walked over to where he was sitting, stuck my finger in his face, and turning to one of my friends, demanded, "Who's this?"

I felt my finger suddenly enveloped by Jerry's very warm hand. Our eyes locked. The first words he ever directed to me were memorable. He said, "Get your finger out of my face, lady!" And I've been with him ever since.

Except for our divorce.

LOVE AND MARRIAGE

Jerry and I had been married twelve years when my father died unexpectedly at the age of seventy. My mother had died of cancer ten years earlier, and now I was an orphan. Within a year of my father's death, Jerry and I were divorced.

I've often thought about what caused our marriage to come apart, despite the love we had for each other. After years of reflection, I believe I now understand what happened. Jerry came to our marriage with three children and certain guilts about not being there enough, although the kids spent a great deal of time with us. When we first met, he made it very clear that his primary responsibility was to his children. Women like to feel as if they're the most important thing to their mates. Our arrangement was fine for me, because I got any extra emotional stroking I needed from my father. It was the extra bonus I received for being his only daughter, his first and only little girl. He was always there as a bolster, supporting me through the best and worst of times. He provided me with an unconditional love that made me feel secure. His death devastated me. My response was drastic. I decided to change the rules of our marriage. Everything that had been acceptable before became unacceptable now. I wanted to feel taken care of by Jerry. I wanted him to be as thoughtful of my feelings as I'd always been of his. I wanted him to change, and show me he loved me as much as he did his children. I wanted him to remember my birthday with the same facility with which he remembered football scores. I wanted him to remember our anniversary with the same total recall he showed for every case he'd ever tried. I wanted him to plan a romantic dinner at a fancy restaurant on Valentine's Day every few years. It was no longer okay for him to hand me a gift as an afterthought a week after my birthday. No more bouquets the day *after* our anniversary. He was able to remember all the children's birthdays; why couldn't he remember mine? I was jealous. I needed attention.

It was war. War is hell. We both became very unhappy very quickly. I was relentless, and Jerry was angry. I had changed the rules after twelve years without warning. Isn't it amazing that I can tell you all about it now, but in the midst of it, I don't think I could have told you exactly what happened? It wasn't that I was wrong to present him with a shopping list of my revised needs. But his response was understandable. "You've changed the rules, and I don't want to play." We separated—and divorced. There was never really a point that we didn't speak to each other and see each other, because we had children having children by this time. We traveled to Massachusetts together for the birth of our second grandchild. Still, Jerry was angry with me, because he felt that I'd broken up the family. And the truth is, he didn't function all that well without me. I handled so much of the day-to-day stuff of couples—laundry, his shirts, dry cleaning, social and medical appointments— you get the picture.

For the first six months we were apart, I enjoyed all the things I'd missed. I went to see only movies *I* wanted to see—not one war picture. I went to the gym when it was most convenient for me. I had a manicure when I wanted a manicure. I did what I wanted when I wanted to. I saw all the children constantly. Eventually, I even dated a little. I realized that I could get along just fine on my own. The question was, did I want to?

Other men left me cold. They all had their own routines down pat—every guy had his own personal *schtick*. At least Jerry had hair.

I began to think that maybe I was asking far too much

from Jerry. I had to face some hard truths about myself—my desire to be loved beyond all reason, and the heavy burden my demands had placed on him.

Women are nurturers. Men are warriors. I needed some extra nurturing. Don't just tell me you love me. Show me! But Jerry was Jerry. There were certain things about him that weren't going to change. The point was—as long as I knew he was trying, that he was always in there pitching, would that be enough for me? And then there was that intangible something I still felt for him, the same thing I felt the first time I ever laid eyes on him. Jerry has a click; he has a sizzle. He's a loving father and grandfather, and he has such a good sense of humor. He'd have to. I love him, and I know that he loves me. We remarried within a year.

There are times to defer and times not to defer. You have to do what makes you happy, but I learned the hard way that sometimes what you think will make you happy won't. You have to step back from your emotions and decide what matters the most.

Just a word about our second wedding. It was performed in the chambers of Jerry's former law partner, now a judge. When Jerry's buddy reached the point in the ceremony when he asked me whether I took Jerry for better or for worse, I stopped him cold, looked Jerry dead in the eye, and said, "I take you for better or *forget it*!"

He believed me.

CHAPTER 2

What Goes Up Must Come Down

My father used to tell a joke about a little boy and a little girl.

BOY: "I can hit a baseball *one hundred* yards!"

GIRL: "Well, my mommy says I can cook dinner as good as an adult."

BOY: "That's nothing! I can lift *twenty-five* pounds."

GIRL: "My mommy says I'm the best tap dancer she ever saw."

BOY: (Pointing to his crotch) "My daddy says I'm better 'cuz I got one of *these*."

GIRL: (Pointing to *her* crotch) "Well, my mommy says I got one of *these,* and with one of *these,* I can get as many of *those* as I want."

The joke is a bit dated now, but it was meant to be cute. Now I think about that joke, and I realize that there are a great many girls and women who believe that if they get as many of "those" as they can, they'll have power. Well, let me tell you something. It's a lie. Using sex to get a man does not give you power—and it probably doesn't give you much satisfaction, either. If you doubt me, ask yourself this: How much power does a young woman like Monica Lewinsky have as a result of her affair with the President of the United States?

Does she have more options, or fewer?

Does she have more legitimate opportunities, or merely infamy?

Does she feel better about herself, or worse?

Is she happier and more confident, or is she sadder and more confused?

I think the answer is clear.

I have often heard the complaint that "men use sex to control women." I don't think that's true. Rather, women allow themselves to be controlled by their own desire to please, not to make waves, to avoid confrontations, to make things easy and comfortable for men, to be more popular, to be elevated socially, to be taken care of financially—the list goes on. The next time you're tempted to accuse a man of sexual control, think about that.

I have to admit, though, that it's very easy for women to cater to men when it comes to sex. For a great many men, their sexual prowess, their perceived virility, can be the be-all and the end-all of their existence.

THE PHALLUS FALLACY

Shortly after Pfizer Pharmaceutical introduced the male potency drug Viagra, I was a guest on *Larry King Live*. Of course, Larry wanted to know what I thought about it. I was amused. He wanted *my* opinion? The last time I checked, I wasn't an authority on male sexual dysfunction, but Larry guessed—correctly—that I had an opinion. In my defense, *he* asked.

I said, "It amazes me that all this fuss is being made over a pill for a couple of guys who are impotent."

Larry performed his trademark double-take. "A couple of guys? I heard that 270,000 prescriptions were written for Viagra the first week it was on the market."

"Larry," I said, "they haven't written 270,000 prescriptions for 270,000 guys who are impotent. Those prescriptions have been written for guys who are hoping for an extra boost. They rationalize, 'Well, if I get some extra, then she gets some extra, too.' This is what I call the phallus fallacy."

The day after the show, half the friends who called me said, "Way to go." The other half didn't mention my remark, and that didn't surprise me. On the scale of dirty little secrets women don't talk about, sex is right on top.

Look, most women know, whether they're willing to admit it or not, that most men believe that all they have to do to satisfy a woman is to have an erection. They think Viagra is the Holy Grail.

I'm not, of course, talking about genuine cases of male impotence. That's not really what all the excitement is about. Men are thinking if they have enhanced sexual pleasure, so will their partners. One more reason to skip the foreplay.

Men should be careful about what they wish for when it comes to technological miracles. I've heard it said that thanks to the advances that have been made in modern technology, men are basically replaceable. There is *in vitro* fertilization. There are vibrators. And they finally made luggage with wheels.

That's my take on Viagra—except to say one more thing:

A Cure for . . . ?

How much money was spent on developing Viagra? Millions of dollars? To paraphrase one of the scientists interviewed about that question, "No, no. Actually, we were testing the Viagra medication as a treatment for high blood pressure, and we happened to notice that, as a side effect, the drug produced instant erections. Durable, too, with a shortened refractory period. Hallelujah! What else could we think?"

What's the biggest threat to the population we face—apart from one another? What disease is the biggest killer? Lack of erections?

Cancer. Heart disease. Lung disease. Diabetes. Cystic fibrosis. Multiple sclerosis. AIDS. Where did the money for the funding of enhanced erections come from?

"He Tries . . ."

So back to the dirty little secret. Often, men and women are having sex with each other for entirely different reasons. Women will have sex because it feels good, it feels warm and loving and intimate. Men will have sex for the purpose of having sex. To quote Freud, "Sometimes a cigar is just a cigar."

Men don't need to be in the mood. They can be ready to go while standing there drinking a cup of coffee and chewing on a half-eaten piece of toast. Men can be aroused from nothing to something in the time it took you to read this. Women—to be satisfied—need the mood. If you're seething with anger and frustration over yet something else your mate did that you specifically asked him not to do, if the kids are crying in the background, if the plumber is on his way to fix a leak in ten minutes, that mood is shot. Oh, and it helps to know your mate is intent on ensuring your physical release somewhere in the vicinity of his own—or at least in the same sitting.

When I've asked women about their experience with sex, I hear one phrase all the time—usually uttered with long-suffering, tender understanding:

"He tries."

He *tries*? What does *that* mean? And the women launch into apologetic explanations about how hard it is for their mates to get them in the mood, to make them feel sexual, to give them pleasure.

I can't help imagining the reverse scenario. How do you suppose a man would react if a woman, once satisfied, turned over and started snoring before *he* reached orgasm? Would he say tenderly, "Well, she *tried*?" Not in a million years.

So here's the basic inequity again: Men can have sex when they're glad and when they're mad; they can have sex after a fight over money, or a battle over the kids. Men can have sex when they're tired, sick, and haven't bathed for days. They don't care.

Women? You can't be yelling at a woman one second and making love with her the next and expect that she's going to experience any pleasure. You can't catch a woman at midnight after she's been up since 5 A.M. doing the million things it takes to keep life going in your domicile, and expect an instant mood. It doesn't work that way.

Which is why, in the end, women don't experience as much pleasure from sex as men do. Men get off every time. Women—well, it depends.

A story: It was an intimate sexual moment. He asked breathlessly, "Now?"

"No," she replied. In a millisecond it was over, and she sighed. "What part of *no* didn't you understand?"

SEX AND THE SINGLE TEENAGER

We live in a sex-obsessed culture, so it shouldn't surprise us that so many teenagers are sexually active. What else can we expect when we feed them a steady media diet of sex-starved hunks and amorous babes? Kids have very good instincts for BS, and they've figured out that all our talk about just saying no and saving yourself for marriage doesn't square with the way we're actually living. So, now that we've created this dilemma, how do we convince our teenage daughters that they shouldn't be having sex?

We can talk to our daughters about what's moral. We can tell them what's right. We can talk about responsibility. These are all perfectly necessary messages. But I think what we really need to impress upon our daughters is that, for a teenage girl, sex is *stupid*.

Even the most esteemed sexual commentator, the fabulous Dr. Ruth Westheimer, would have to agree that a fifteen-year-old boy will not give the time, effort, or energy—nor does he have the knowledge—to satisfy a sexual partner. He can satisfy himself; he's had practice. But a partner? Never.

Teenage girls have to recognize that what they will get out of sex is probably not pleasure. It's definitely not love. What they get, in most cases, is social acceptance, a Saturday night date, the chance to be a member of the popular crowd.

Maybe they'll also get a sexually transmitted disease. Or they'll get pregnant. And then their entire lives are changed forever, and not for the better. Is it fair? Maybe not, but it's a fact.

I have always said Madison Avenue is very strong. They

can get you to buckle up for safety; they can convince you not to drink and drive. Why can't Madison Avenue say, in a way that sticks, that sex for teenage girls is dumb?

By the way, I do not believe in setting up nurseries in schools so that teenage mothers can continue their education. I'm sure I'll get a lot of flack for this, because we all want girls to have schooling, even if they get pregnant. Here's my point. Once we make it easy, normal, acceptable, to be a teenage mother, we've lost the battle. I have a better idea: Let's communicate to our daughters that when a boy says, "Show me you love me, have my baby," it's the dumbest statement ever uttered. And it would be a true embarrassment if she actually bought this pathetic line. Let's ask ourselves what we can do to guarantee that our daughters know the truth about sex before they even hit puberty. "Just say no" doesn't always work. In fact, sometimes curiosity kills the cat.

THE MORAL EQUATION

While we're on the subject of sex, I want to say a few words about our current moral climate. Mind you, I don't believe morality and sex are synonymous. I've seen too many examples of immorality while I sat on the bench that had nothing to do with sex—such as a parent abusing a child, or a mother giving birth to a baby addicted to crack. These days, it seems, discussions about morality usually focus on sex. And in this area alone, I certainly think that the line between what's right and what's wrong has been blurred.

Recently, I watched a television biography of Loretta Young, and it dwelt on the suggestion that she had had Clark Gable's child. What amazed me was the elaborate scheme she allegedly undertook to hide this fact from the world. At that time, if a movie star had an illegitimate child, it was enough to destroy her career. And the studios were eager accomplices, spending big bucks to make sure the truth never got out.

Today, it's a very different story. Movie stars, our heroes, are often seen having two or three children without benefit of marriage, and almost casually moving from one mate to another. No one bats an eyelash; in fact, it almost enhances their careers.

I think that is a lousy way for heroes to behave. But I am also realistic. I know that some things have evolved, and times have changed.

Jerry and I had been inseparable for one and a half years when I decided it was time for him to make a commitment. He was not unhappy with the status quo, but was quite happy to move in with me (saving that late-night travel time).

I said, "Marriage, not living together."

He said, "It's just a piece of paper. Why should we let the State control our living arrangements?" and other familiar arguments from the male front. He also pointed out that I was thirty-two years old, I had two children, and no one would blink an eye at a live-in relationship.

How to counter? Must think fast, girl.

Jerry adored my father. He also knew my father would eviscerate anyone who hurt me. So I said, "Living together is okay with me—if it's okay with my father. *You* ask him."

We were married three months later. That was then. It wasn't that long ago when my daughter was engaged to be married, and she was living with her intended. Two weeks before the wedding, they were planning to come to New York to get their marriage license, and they would be staying with us for a couple of days. The day before they were scheduled to arrive, Jerry asked, "Are they planning to sleep in the same room?"

"Jerry," I replied, "they've been living together for the past year."

He set his jaw. "They're not married. They won't sleep together in *my* house."

I could see the storm clouds forming for a major confrontation, so when the kids arrived, I quietly took them aside. "Look," I said to my daughter, "you know Jerry. He's old-fashioned. He doesn't want you sleeping together in the house. So here's your choice. You can go to a hotel, or I can marry you right now."

And that's how it came about that on a Saturday afternoon, I put my robe on over my T-shirt and shorts, asked a neighbor to stand as a witness, and I married them.

"There," I said to Jerry. "They're man and wife. They can sleep together."

A few years have passed, and today all the kids live together, and we pretty much accept it. Is it a good thing? It's probably not such a bad idea if they are emotionally committed to each other. It's another story when people jump from bed to bed, change partners on a whim, and have babies without worrying whether there's a father around.

We can't put the genie back in the bottle and wish people behaved (or at least *appeared* to behave) as they did thirty years ago. What we can do is teach our children the value of responsibility and the meaning of respect, for themselves and for their partners.

CHAPTER 3

Denial Is a River in Egypt

What is denial?

Denial is when your fiancé asks you for a loan so he can buy you an engagement ring, and you give it to him.

Denial is believing your married lover when he says, "My wife and I haven't had sex for eight years," even though they have a two-year-old child.

Denial is thinking you'll advance your career by working long hours for little pay, while your boss takes credit for all your good ideas.

Denial is when the man you love tells you that he

doesn't want children, but you marry him anyway because you just *know* he'll change his mind.

Denial is believing that your adult children and their spouses will appreciate your setting them straight on everything from where they should live to what they should eat.

Denial is buying a dress that is two sizes too small because you're planning to lose weight.

Denial is thinking that having a baby will strengthen your rocky marriage.

Denial is believing he can't give you his phone number or address, and can't see you on weekends and holidays, because he's a top-secret agent for the CIA.

Denial is making excuses: "He only lied once."

Denial is saying, "He only does a little cocaine."

Denial is saying, "He tries."

Denial is saying, "He only hit me once."

Denial is believing that he'll respect you in the morning.

Denial is an insidious ailment. It starts when you're young and takes root, if you let it. In the beginning, the state of denial places you in a comfort zone—what you don't see can't hurt you. In time, reality seeps through the walls, and it feels like a kick in the teeth. If you have ever woken up one morning and asked, "How could I have been so blind?" you know what I'm talking about.

WHAT WE DO FOR LOVE

How many men do you know who would give up a terrific job to follow a woman halfway across the country? None. But women do it all the time. I know about this, because I did it myself. My first year of law school, I attended the Washington College of Law, American University, in Washington, D.C. There were 126 students in my entering class that year, and I was the only woman.

I loved law school. At the end of the first year, my grade point average placed me number one in the class. I knew I had a bright future. I was hooked on the law, and I enjoyed the respect of my professors and classmates.

During that time, I met my first husband, who lived in New York. He was already a lawyer, working for little pay at the Legal Aid Society. He asked me to marry him while I was in my first year of law school, and, with stars in my eyes, I accepted. There was no question that by accepting his proposal, I was tacitly agreeing to transfer to a law school in New York, because that's where he worked. I transferred, and the remainder of my law school career was unremarkable.

So when I say that women do dumb things, I speak from experience. In retrospect, leaving a law school where I was the number one student in my class was a dumb thing for me to have done. Think of the options I didn't even consider. I could have waited to marry until I was finished with law school. Or my husband could have taken a similar position in Washington, D.C. (Let's face it, lawyers aren't exactly beating a path to the door of the Legal Aid Society.) But neither of us gave any thought to such possibilities. I automatically deferred to my new husband, unconsciously and without question.

For many couples, things are more equal today, but for the vast majority it's probably much the same. I've often wondered how my life would have been had I finished law school in Washington, D.C., graduating at the top of my class. Would I have gone on to the Attorney General's office or the Justice Department—the positions top law school graduates are slated for in Washington? How different might my entire life have been?

Of course, whenever you start playing out the possibilities, the roads not taken in life, speculation becomes futile. Perhaps it was a favor to me. I got a couple of wonderful children out of the deal. When I grew up, I thought, God, that was really foolish. I mean, not even to have a discussion about what would be best for both of our careers and lives in the long term, but instead just to pick up and move myself back to New York to be with my new husband—without question! It still amazes me that I did that.

What's Your Breaking Point?

Men are far more unlikely to stay in an intolerable situation than women. If they do stay in a bad marriage "for the sake of the children," it means they're getting their satisfaction elsewhere. It may not be a sexual affair, necessarily. It may just be that their energies go into their work, their golf games, everything but you, everywhere but home. Women will stay in intolerable situations for any number of reasons, but primarily because they fear being left alone. If the emotional core of a relationship is missing, most women feel as if they have very little, if anything. Even though the relationship may otherwise provide many levels of comfort and security, financial and otherwise, all else withers and turns gray without the emotional element. And yet too many women will stay, ultimately feeling that any relationship is better than none at all. Having said this, it's a fact that women survive being alone much better than men do. Men, who've been coddled their entire lives, suddenly find themselves having to grocery shop, do their laundry, clean up after themselves. It can be a terrible shock. Most men will stay in a relationship if their basic needs are met. If they're fed, allowed to watch all the TV sports they want, and occasionally get sex, they'll stay forever. The truth—what most men really want is for someone to take care of them. That being accomplished, the ancillary annoyances, frustrations, rages, and disappointments all become as the roaring of great apes, the barking of dogs. Accept it for what it is—grousing. Food, clean clothes, TV, a tolerable amount of grousing, food, sex, TV, sex—you've got a man forever. But what about you? Let's talk about *your* happiness. If you expect to be happy in any relationship one hundred percent of the time—dream on. A more realistic expectation is in the neighborhood of fifty-one percent happiness, forty-nine percent stress.

If you can be happy in a relationship more than half the time, you're probably way ahead of the curve.

However, if you think your relationship is not working, and you're considering a breakup, stay realistic about that, too. You have to say, "I'll be happier alone." Be sure that's what you're really thinking, not "I'll be happier with someone else." If you're truly unhappy, you'd better hope that you have the self-confidence and the resources to go out on your own.

When you leave a relationship or a marriage, you need to be able to do so in a spirit of self-esteem. If your grievance is that your mate isn't as wonderful as another man might be, you're running on the same treadmill.

EMPTY THREATS

If the man you're with is threatened by your achievements or abilities, get out. He's not going to turn around and suddenly be supportive. It's unlikely he'll change. Quite the opposite, in fact. Insecure men will do whatever is necessary to bring you down—especially if you threaten them with your intelligence and success.

I poke a lot of fun at my husband in this book, but we love each other. He revels in my success, and I in his. He couldn't be happier about my new television career. Jerry is a judge in the Supreme Court, Bronx County, the City of New York. He handles serious cases—murder, rape, robbery—the worst. He's a tough, fair, honest man, and he doesn't allow any crap in his courtroom. He also happens to be a wonderful teacher, and so, over the years, junior high and high school classes have visited his courtroom to watch the proceedings. Jerry meets with them in the courtroom during the lunch break and discusses the particulars of the case they viewed that morning.

Recently, he had a class of predominantly female high school juniors, about a dozen of them, sitting in on a brutal and complex murder case. At the lunch break, Jerry patiently explained the details of the case and described what they had witnessed that morning. Finally, he looked around and asked if there were any questions. Every hand shot up in the air. Wow, Jerry thought, what eager, interested students! He was impressed. He nodded to a dark-haired young girl standing to his left and asked her what her question concerned. She beamed. He had picked her first.

"Are you really Judge Judy's husband?" she asked excitedly.

Jerry was initially taken aback. He hadn't expected that, but he was a good sport. "Yes, I am." He smiled. "Any other questions?" Another flurry of hands. Jerry pointed to a second girl.

"Does Judge Judy film her TV show in New York or California?" she asked earnestly. Well, there it was. "In California. She films her show in California," he replied, giving in to the weight of historical imperative and the undaunted curiosity of the young. Finally, two of the girls begged him to autograph their notebooks. He was flattered and happily signed the books for them. But when he returned the signed books to them, he noticed the girls seemed unhappy.

"What's the problem?" he asked solicitously.

"Couldn't you just sign it 'Judge Judy's husband'?" one of the girls said.

Consider the men you know. How would they react in a similar situation? I'll tell you what Jerry did. He came home and reenacted the scene for me, until we were both laughing. He's my biggest cheerleader, my biggest fan, and it's always been this way. He's capable of doing that because he's solid and secure in who he is. Jerry's ego isn't threatened by my success—it's enhanced.

THE GAMES PEOPLE PLAY

Every couple develops its own particular quirks in a relationship as the individual personalities and values clash, combine, and eventually merge. People have broken up because of ice cubes—literally. During my first marriage, a young couple who lived in the same Brooklyn apartment building separated because of ice cubes! They had the same modestly sized refrigerator we did, with the same small freezer compartment. Apparently, they both used ice cubes in everything they drank—and apparently they both drank a lot of iced tea. The clash was over refilling the ice cube trays and marshaling extra bags of ice cubes so that there was always an ample supply. She was a champion refiller and surplus ice cube marshal; he was a champion user of ice cubes with not a thought of resupply in his head.

I witnessed them clash over this point repeatedly. I heard them arguing about it when I passed their apartment on my way home from law school some evenings. It finally occurred to me that if they were constantly arguing over this one issue, what other issues lay unresolved between them? It was really about two different kinds of viewpoints, two people unable to see each other's sense of values, and so of course it was about much more than ice cubes. They separated within the first year of their marriage and eventually divorced.

HIDING SHOES

A wonderful female physician, very successful, loves to buy shoes, but her husband, also a prominent physician, goes ballistic whenever she does. I happened to bump into her at a shoe store in the city, where she'd just purchased a couple of gorgeous new pairs of shoes, and we started chatting away, only to discover that we were both going to be at the same charity event later on that evening. She suddenly looked alarmed. She leaned in and said, almost in a whisper, "Look, Judy, if we happen to see each other tonight, don't mention anything about our running into each other here earlier, okay? I'll be with my husband, and if he hears I bought new shoes, it'll spoil the entire evening. He gets crazy when I buy new shoes. I throw away the boxes and just slide them into my shoe closet, so he never knows. Promise me. Our little secret."

What was I going to do, turn her in? I agreed. Of course I would keep her secret. But that she allowed herself to be intimidated and controlled really ticked me off.

What is this about? Why on earth would a grown woman, perfectly capable and with her own income, hide shoes from her husband?

I have a friend who has been married for thirty-five years. Claire still receives an allowance each week from her husband. An allowance!

"It's funny, isn't it, Judy?" she said. "I have unlimited use of all the credit cards, but Herman insists that I never have more than $150 in cash each week."

This revelation came one day when we were together and Claire had to purchase something in cash. The place didn't take credit cards. She was embarrassed but shrugged it off. "Herman's just a little quirky about money, I guess."

He's not quirky. He's controlling, and Claire allows herself to be controlled. Remember, the ground rules you establish at the beginning of a marriage are very hard to change.

THE INFANT FORMULA

This is so classic I call it the Infant Formula. In how many marriages do you see women having babies to keep the relationship together? Things going bad in your marriage? What's the answer? Have another baby to bring it together.

You know that bloated, sick feeling you get after you eat a heavy meal?

You're really feeling uncomfortable. Then they bring around dessert, and you begin to rouse yourself. As long as you're feeling so lousy anyway, why not have a big piece of pie? That sounds silly, doesn't it? Believe me, deciding to have a baby to save a miserable marriage is no different.

It's especially crazy when you consider that in most marriages, one of the principal elements of frustration is dealing with the children. The constant responsibility of having to deal with the kids can be a deal-buster. "Ever since we had Tatiana, you haven't paid any attention to me. All you think about is her! What about me?" The answer to the problem? Another baby. The period of pregnancy is usually a euphoric time in a marriage. The planning, the excitement, the anticipation, can fill some of the empty spaces. But then, inevitably, the baby arrives, and you're right back in the doldrums—times two. The decisions we make as young people aren't without consequences. Once you have a child, you can't say, "I'm going to move on now, and just put a period to that part of my life." Believe me, I'm all for getting on with your life when a relationship ends, but children permanently alter

the equation. Having babies is a serious business with last-
ing consequences. I wonder why it's still used as a way of
shoring up crumbling relationships?

I think it's a troubling combination of denial and fantasy wishes. First, denial that the relationship is in serious
trouble, and second, the fantasy wish that having a baby
will dissolve all the current tensions, and all the problems
will go away. There are women who actually believe that
with a new child to love and care for, the marriage will be
even stronger than it was before. Perhaps it's another
example of how painful it is to accept that a relationship
has failed—so painful that many people automatically shift
into the denial mode. A baby will make it all better.

Whose Money Is It?

Another form of denial has to do with money. A woman
may think she is financially secure because her husband is
bringing home a good paycheck, but that isn't necessarily
the case.

Ask yourself: Do you have control of your own money?
Your *own*, not *his*. It's an important distinction. Even when
a woman controls the money, and her husband gives her
the paycheck, it remains in his power to cut her off. A
woman may pay the bills, she may appear to control the

money, she may decide what to spend it on, but the reality is, if the man leaves, so does the paycheck, and suddenly there's nothing to manage.

A woman I know discovered that her husband had been having an affair, and she threw him out of the house. A few days later, when she went to the grocery store, she discovered that her credit card had been canceled. She called the bank and found that all her accounts had been closed. On his lawyer's advice, her husband had shut off all her access to money. She was helpless. She couldn't pay the bills, she couldn't feed the kids, she didn't even have money to buy gas for the car! She capitulated, and they got back together.

Have a profession, a vocation, your own source of money. It's your security, girls. Nobody can take it away. It sets you free emotionally. Every woman has the ability to earn a living—preferably at something she enjoys doing. Sometimes you might be forced to accept a less-satisfying position. My first job out of law school was a real bummer. But keep your eyes open, prepare yourself, and network. Seek out people in your area of interest, take a course, attend a seminar. Leave the house. No one is going to stumble over you if you're sitting in your living room watching the *Judge Judy* program on TV!

DIVORCE 101

Wising up is hard to do. If you start a relationship denying your needs and desires, it's a shock when you realize that you've been had. You may be boiling with anger, enraged. Suddenly you've transformed from the agreeable, compliant mate and become a ferocious creature your husband doesn't recognize. Now you're headed for a truly acrimonious divorce.

In most marriages, it's the woman who smooths things over, defers, makes everything nice. Whenever there's a fight, it's usually the woman who ends up making up. Why? If the woman is at home with the children, she's got all day to think about what happened, to let it fester and gnaw at her. "I should have said this when he said that," or "I should just get out of here. I can't stand it anymore." It ruins her entire day.

And the participating mate? The arguer to her arguee? Well, he's gone off to work, and he's doing his thing. If he's thought about it at all, it's only peripherally in a busy day. The two of you had a fight, he walked out and probably had a grand day at work. He walks in the door that night, it's just the same old, same old to him. He's already forgotten about the fight. So, to keep the peace, the woman lets it go. But it's building inside. When she reaches the point of talking divorce, she comes out swinging—and the man thinks, "Who is this angry person? Where did all of this come from?"

It came, my boy, from years of denying her own needs, of subjugating her desires for the sake of the family, of denying her dreams for his realities. I practiced matrimonial law for a while—divorces, prenuptial agreements, separations, orders of protection, the entire gamut. My clients would all chant the same basic mantra—"I want to get the bastard. I want everything! I want him to suffer. I'm going to do *this* to him, and then I'm going to do *that* to him. I want to make him pay for what he's put me through. I want to break him!" The message was always: "You're going to pay for this for the rest of your miserable life. I don't care whether you're happy, content, or can even afford to live. All I care is that you support our three children until they turn eighteen years of age. Live in a hole in the ground. I don't care."

How do you think most men react to this kind of attack and pillage mode? Not well. Men are conditioned to respond badly to what they take as threats. They automatically go into their warrior stance—it's almost a reflex reaction. That's what causes most of the acrimony in divorce cases. Instead of being reasonable and mature and seeking a realistic settlement, couples lapse into hostile negotiations that bring out the worst in everybody.

When women came to me with fire in their eyes and

revenge in their hearts, I would call my Divorce 101 class to order. "You've got it all wrong," I'd begin. I could tell they didn't believe me. They were sure they had it exactly right. "Men don't deal well with guilt and anger," I explained.

"You attack them, the majority of the time their reaction will be to fight back. Men are combative. It's their nature. So if you come in pugnaciously and vindictively, aching for a fight, that's what you'll get. Most men understand that. Once you're the enemy, the opponent, they're going to fight. Think tactically. Don't get angry or combative—instead, get what you want. Here's how."

And then I would tell them: "You are the victim, the wounded party, the little bird fallen from her safe nest. Let them know you're scared, frightened, you don't really know how you're going to be able to get by on your own."

Of course, many women bridled at my advice. Show I'm scared? Never! I always patiently explained that this was merely a negotiating stance, a tactic. I counseled them that if they entered into their divorce negotiations in a combative manner, they could expect a bitter fight. Men are natural warriors.

There's another reason. If you don't attack, many men will have to face the hard reality that they have taken more from your relationship than they have given. You've denied yourself for the sake of the family, and if they now find themselves in love with their twenty-year-old assistant, it's their crisis, and not of your making.

GUARANTEED DIVORCE INSURANCE

If a woman has an education and a skill, she has guaranteed divorce insurance for her lifetime. And it's important to have the ability to do something! Let's be clear that many relationships—or people's ideas of what relationships are supposed to be—are based on whims, fantasies, and unrealistic expectations. Young women place themselves in these roles, wrapping themselves in twisted romantic fantasies, fairy tales.

The reality is much harsher and definitely more mundane. And since half of all marriages don't survive, women had better shake themselves loose from their fantasies and get divorce insurance.

What is divorce insurance? It's an education, a skill, a profession, an ability to take care of yourself, come what may. Divorce insurance may seem shocking, even cynical, but look at it this way. Everyone has homeowner's insurance, yet the likelihood of your house burning down or being felled by a storm is statistically small. However, the likelihood of disaster striking your marriage is fifty-fifty.

The only insurance protection you have is financial

independence. It isn't enough to marry a rich man. Should you feel comfortable with fifty percent of his profits signed over to you? Not on your life. How many successful businesses have you heard about that mysteriously stop making profits as soon as divorce enters the equation?

You have to be smart, use your head, plan. Hope that everything will remain as wonderful as it was on the day that you were married, but be prepared in the event that it doesn't. The only insurance you can get right now—I don't think anybody is writing divorce insurance—is a vocation, a way for you to earn a living that you enjoy, that has some meaning for you, and that makes you independent and self-supporting.

Choose something you're good at, something that you look forward to doing each day. Whatever it is you choose, make an effort. The more you give of yourself, the more you end up getting back. Becoming a fully integrated human being can happen at any time in your life, no matter your age or level of experience. That's the best insurance there is.

CLIMB EVERY MOUNTAIN

My first marriage taught me what happens when a woman denies her needs, denies what she wants, denies what gives her pleasure, to be a cooperative partner. But apparently I hadn't learned the lesson well enough when I met Jerry Sheindlin.

Jerry loved to ski, and he was a wonderful skier. I listened to him sing the praises of racing down icy mountains with the sleet and snow blowing in your face, and I was skeptical. Frankly, it didn't seem all that appealing. In the interests of trying to accommodate this man to whom I was wildly attracted, I agreed to try his favorite sport. I outfitted myself in the requisite parka, ski pants, boots, and gloves, and one weekend we headed for the mountains.

By the time we boarded the rickety chair lift and started on our way up the mountain, I was already scared to death. I hung on for dear life as we rose higher and higher. When we came to the top and they informed me I was supposed to jump off, I said, "I'm not getting off this thing." So they pushed me off.

There I stood at the top of the mountain, freezing, terrified, and unable to move. Meanwhile, Jerry was bounding through the snow like a happy puppy. I looked down and realized that short of calling for an airlift, there was only one way I was going to get to the bottom. So, slowly, very slowly, we crawled down.

We headed for the lodge for dry clothes and a hot drink, and I decided we had to have a little chat. I had promised Jerry that I would ski with him because he loved it so much, but having tried it once, I was ready to face the truth. This wasn't a time for denial.

"Jerry," I said to him, "you can have a *screwer* or you can have a *skier*, but in this woman you're not going to get both."

He stared at me openmouthed. I added, "But I remind you, you can only ski in the winter."

I never set foot on another mountain.

CHAPTER 4

Master the Game— Then *Play* It

Which women are successful? The ones who can stand toe to toe with men—and do it with humor. Thanks once again to my father, who taught me to have a sense of humor, I learned that lesson early on.

When I was a newly appointed judge in the Bronx back in 1982, the judges' lunchroom was strictly a men only affair. I knew that, and I didn't go in there. I'd eat lunch in my chambers instead. It was made clear that the judges were all enlightened, and that they were certainly prepared to *tolerate* a woman in their lunchroom—it was just that they preferred if there weren't women there to tolerate. I got the idea. There was a day when I needed to speak to one of my colleagues, who happened to be having lunch at a table full of my male brethren. I went into the lunchroom, pulled up a seat at the table, and began to talk with him. While we were engaged in this conversation, an older judge approached me at the table and said, very officiously, "Excuse me, young lady, but this is the judges' lunchroom."

I know I could have turned to this judge and quietly said, "I'm Judy Sheindlin. I'm a judge in family court," but I didn't. Instead, I rose from my chair with my eyes downcast and said almost in a whisper, "Oh, excuse me. I'm sorry. I'm just here to clean up." Then I began wiping crumbs off the table and gathering up dishes. The officious judge stood there watching me with a smug, satisfied look on his face, his arms folded over his substantial belly. The other judges at the table, who knew who I was, were taken aback by the scene, but they were caught between mortification and delight at my reaction to his behavior. After gathering up the dishes at the table, I left the lunchroom. I returned to my chambers. Five minutes later, the officious old judge came storming in, furious. "Why didn't you tell me you were a judge?" he shouted. "You embarrassed me!"

I embarrassed *him*. He acted like an imperious jerk, and I embarrassed him. I got a big laugh out of that.

Why did I respond that way? I was playing with him—pretending to be the maid—because he assumed I couldn't have been a judge. I guarantee you, he was more careful the next time he started to make an assumption about a woman showing up on male turf. And I enjoyed the little scene immensely. It made my day.

KEEP A HOOVER FILE

Knowledge is power.

I believe in the power of the individual. I believe that one determined, skilled person can do just about anything. That's why I keep a Hoover File, as in J. Edgar.

Men still control the workplace for the most part. My experience has been that the law, the courts, and the television business are all male-dominated. The higher a woman rises, the more people come out of the woodwork wanting to knock her off her high horse.

Being well aware of this by the time I was appointed as a judge in the family court, I decided that it would be in my best interests to keep what I came to call my Hoover File. The Hoover File is an information file—whenever I heard anything about someone, I'd write it down and file it away for future reference. In my twenty-five years in family court, I only had to pull out my Hoover File once. A particularly sensitive case had come before me. There's no need for specifics, but suffice it to say that some very powerful, well-connected people in the city wanted the case disposed of in such a way as to cause no embarrassment to them or the institutions involved. It was a custody case in which the city's child welfare agency made a series of tragic errors that deprived a mother and son from being together for a very long time. The details were bruising and pathetic, showing a trail of incompetence and malfeasance that was shocking. This case had been kicking around like a hot potato for years until it landed on my bench.

I was in my chambers early one morning before court, when a colleague of mine stopped in to see me. He said, "There was a breakfast meeting this morning, and your name came up."

"I'm flattered," I replied, with the barest hint of sarcasm. "The power brokers were talking about me."

He shifted uncomfortably. "You're up for reappointment in nine months, you know. You're a good judge. Don't let this case destroy you."

Did I detect a subtle threat? It sure *smelled* like a threat. This was very interesting. I was an outspoken judge, but I was good. I worked hard, and I was rarely reversed in the appellate courts. Until that moment, I hadn't even considered that I'd have any problem being reappointed.

I stared at my colleague, the messenger from on high, as if he were a bug that had splattered on my windshield. Then I stood up and walked over to a file cabinet and pulled out a drawer. I pointed to a row of files. "It's amazing how much information I have collected over the years on your breakfast buddies," I said. "*They're* discussing me? You'd think they'd be more concerned with my discussing *them!*"

My colleague just shook his head and headed for the door. "Wait," I said. "Do me a favor, will you? Tell the power brokers that I don't expect to have a problem with reappointment because of my handling of this particularly pathetic case. In fact, I expect to sail through the reappointment committee. If it causes them embarrassment, that's *their* problem."

I put my Hoover File in my briefcase and brought it to court with me that day. And to lunch. And home. I felt as though I were carrying my guarantee of my continued employment in family court around with me—and it was heavy. Months later, I began the reappointment process. I walked into the committee, shook hands with everyone, and sat down. The committee chairman said, "How does it feel to be a legend before you're even fifty?" I grinned. "It feels terrific. I like it!" Rising from his chair, he said, "Thank you very much for coming in to see us—we're done." And that was it—in and out. Jerry was standing outside waiting for me with another judge who was scheduled right after me. They had expected to be waiting the usual twenty minutes to half an hour, and I walked in and back out in about thirty-five seconds. They just looked at me with their mouths hanging open. Today I'm in a completely different kind of world, although the television business is as male-dominated as the law and the courts were. I think it's probably less of a problem than it used to be, but sometimes I think they're still not used to dealing with smart, aggressive women. Apparently—at least from what I've seen—they're used to controlling women, not dealing with women who take control. So in many ways I'm a different kind of woman than they're used to dealing with.

I still keep a Hoover File for television. I keep notes of meetings, conversations, and business discussions, so I'll be more fully prepared to deal with the very complex business and contract aspects of the work. I don't know if that's what the "big boys" do, but women have to be smarter when they're dealing for themselves. These are very bright, well-informed people, who really know their business, and it behooves me to try to keep up in a relatively new venue. That means I have to master their game.

FIGHT INEQUITY

The inequities between men and women—the societal, business, and relationship inequities that have been driving women around the bend for centuries—have finally reached a boiling point in only the last generation or two.

Let me give you a word of advice. If you're trying to balance the scales of justice and equality in all your work relationships, you're going to come up short. No division of responsibilities is ever likely to be equal. It's very possible that you may never get exactly what you deserve in a relationship, in your work, or in your life, for that matter. What *can* you do? Be smart. Use your female intuition, your innate feminine wisdom, and your energy, and be like a sponge. Soak up information and detail, learn every aspect of what you do. Once you've accomplished that, you have to take incremental steps to make sure that the people who are in charge understand that you're ready for more responsibility, that you've made yourself indispensable to the operation of their business. Now that's quite a trick, isn't it? You have to be assertive and aggressive, but do so with a light touch, mixing confidence with humor, and concentrate on your work. The outcome is that eventually you have every bit of information you need to succeed. You'll have established a network of communication with those in charge. The difficult part lies in combining the professional, the social, and the personal interactions into a secure power base.

Realize that you start out with a handicap. Aggressive, assertive women are often viewed as bitches and ball busters, while the same qualities earn men the distinction of being tough, no-nonsense, talented, and good leaders. Mega-merger tycoons like Ronald Perelman, Henry Kravis, Rupert Murdoch, Ted Turner—throw in a few of your own current favorites here—all are considered great and ruthless businessmen. Nobody faults them for what they do or how they behave. Women are too afraid of being called bitches. They want to be liked. Men want to be respected. Don't be afraid of labels. Don't let the insecurities of others scare you away from your goals. Be aggressive—go after what you want. Choose role models among women who are able to combine being kind, warm, loving, and generous with being shrewd, dynamic businesswomen. Look at Oprah Winfrey, probably the most successful woman in the television industry during the last ten years. This is a woman who is obviously in charge of her destiny and her businesses, yet she also has cadres of loving, supportive fans. And she's surrounded herself with an incredible staff of advisers, associates, and employees.

Barbra Streisand, indisputably an enormous singing and acting talent, as well as a gifted film director, is criticized for being a "control freak" because she insists on a certain level of discipline and excellence. On the other hand, James Cameron, who won an Oscar for directing the film *Titanic*, is also extremely difficult and demanding, but he's not called a control freak. He's called "Best Director."

If women are going to continue to be successful, we must toughen our facades, develop a protective shell to shield us from all the flak that flies in our direction. Frankly, if someone calls me a control freak, I automatically say thank you. Don't you think I try to control every aspect I can of *Judge Judy*? I certainly do! The show has my name on it.

BECOME INDISPENSABLE

Whatever it is you do, the only way to be successful is to make yourself indispensable. Learn everything you can about whatever job you decide to tackle. Keep your eyes and ears open, and absorb as much as you possibly can. Let me give you a couple of examples.

When I began doing promotional work for *Judge Judy*, it entailed a lot of traveling around. The syndicator of my TV program, Worldvision Enterprises, assigned a young woman to travel with me as my coordinator and assistant. She was unacceptable. She treated people disrespectfully, as if they were mere props in the background. I didn't want this young woman to coordinate anything, or to represent me. Currently my promotion is handled by a talented young woman named Terri Corcoran, and she is wonderful. Terri is efficient, organized, has a lovely sense of humor, treats people beautifully—we get along tremendously well. She has become indispensable to me. She makes the whole process of doing all of the ancillary things necessary to promote *Judge Judy* possible.

I'm sure there are other people who could do Terri's job, and do it well. The point is, until she came on board, the job hadn't been handled well at all. The difference was striking. So, whenever I have to do promotional work, Terri Corcoran's my woman. Now, of course, she's been promoted to director of station relations for Worldvision. She made herself indispensable. Nancy, the woman who does my hair and makeup for *Judge Judy,* has also become indispensable to me. I'm sure there are a lot of people who could do my makeup, but Nancy has the right kind of personality for me. She's upbeat and fun, tells a good joke, and laughs at mine. When we're done with the makeup and hair, I don't feel drained or antsy. I feel good. So Nancy's become a part of my team, and I find her indispensable.

Now don't confuse being indispensable with being subservient. It's not that at all. It's about demonstrating your worth, no matter what your position—and then using it to move ahead.

My daughter, Jamie, started a new job. Her salary was low, and after a few months, she wanted a raise. She asked me what I thought, and I told her what I'm telling you. "You want more money at a job? Make yourself indispensable. Be the person about whom the boss says, 'What'd we do before you showed up?' That way, when you ask for more money, your boss will think twice before turning you down." It's as simple as that.

LIGHTEN UP!

Women in the workplace have to learn to lighten up, to use humor. When you take yourself and your work dead seriously, you don't give off the aura of confidence. Most women haven't been taught how to disarm a situation with a deft wit or a funny crack. Men do it all the time.

You can't underestimate the power of humor. Fortunately, my father was a great teacher, and I've honed my skills in adulthood.

In my new career as a television personality, I attend an enormous annual programming and syndication convention, called NATPE—National Association of Television Programming Executives. It's where deals are made for the coming year. The first year I attended, I ended up sitting at the *Judge Judy* booth and signing autographs for two hours every day. My hand swelled up, until I got smarter and realized I could control this. So the second year, I informed the powers-that-be that I would sign autographs for only one hour. Still, it's very grueling.

The first day, I sat down at my booth, and I noticed that the three big bosses were standing nearby on the convention floor, having what appeared to be a very intense summit meeting. (Men's meetings always look very intense.)

There was the president of Worldvision Enterprises, John Ryan, who syndicates the *Judge Judy* program; Larry Lyttle, the president of Big Ticket TV; and Peter Bachmann, the president of Spelling Entertainment. At the time, I was in contract negotiations, because the show was a hit. As I was sitting at my booth signing autographs, I decided to tweak the big bosses a bit. People were coming up to me, saying how much they loved the show. So I stopped the next person who approached me for an autograph, and said, "You see those three men over there? Well, if you really like my show, it would mean a lot to me if you would go over there before you leave and tell them. Thanks." She did, and I could see the bosses nodding and smiling—only slightly annoyed that their summit had been interrupted.

Men don't like to be interrupted when they're in a

huddle. That's because everything they talk about is so important—to *them*. Men consider their own time so valuable. Besides, they lose their train of thought rather easily.

All of the above were carefully considered by me when I asked the next person who approached my booth to do the same thing. And the next person, and the person after that. Soon, every person who came to my booth was making a beeline for the bosses. I watched out of the corner of my eye as person after person approached them, interrupting their conversation to tell them, "I really love the *Judge Judy* show!" I wondered which of the three would figure it out first. It was Peter Bachmann of Spelling. After the fourth interruption, I saw him turn and look over at me with a quizzical half-smile on his lips. The interruptions continued. Finally, the big three had to terminate their conference on the convention floor and move to more private quarters off the main floor. After my hour of autograph signing was over, I was about to leave when I was called over to a side room where the bosses had just finished their "meeting."

Larry came over to me and said, "Hey, what's the matter? Don't you like to sign autographs?" Both John Ryan and Peter Bachmann laughed at this. They were being cute.

"I prefer to sign my name immediately following the

words 'For Deposit Only,'" I responded with a smile. They all roared with laughter. Disarming? Yes. Made my point? Yes. What fun! Women have to learn to do that. Don't be afraid to give as good as you get.

A little carefully placed sarcasm never hurts. When the ratings came in for the start of my third season on the air, and they broke all the records, I was elated. I'm sure Larry Lyttle was elated, too, although I couldn't be sure because I never heard from him. Finally, I made a call and left a message on his answering machine. I said that I'd heard about the great numbers, and I added, "Larry, don't pay for the flowers. They never arrived."

I knew how Larry's mind worked. First, he would be confused. He'd think, "Flowers? What flowers?" At length it would dawn on him. Oh, *flowers*.

My father was always telling jokes. Since he was a dentist, he had a captive audience. I listened and picked up his tempo, his rhythm, his take on things, and I think that's a big part of what a sense of humor is—you take an actual life circumstance and see the ridiculous side of it. Then you embellish or exaggerate it a little to make it a funnier story.

As the saying goes, she who laughs, *lasts*.

THE LAST LAUGH

The chief salesperson for Worldvision Enterprises is a man by the name of Bob Raleigh. I'm very fond of him. He's a wonderful salesman. He has a background in psychology. My first year at NATPE, I got out there, and I was selling the *Judge Judy* program with the best of them. They'd bring me in and I would schmooze the station owners and tell them a funny story. It was a very successful strategy—they enjoyed it, and I enjoyed it. The first day there, I asked Bob, "When do we find out how many stations we sold today?"

"We'll have a wrap-up meeting between 6 and 7 P.M. We get all the salespeople together, and we figure out what we sold and what's still open," Bob responded in a polite rote. We'd only met briefly a month or so before, so I could see that he didn't fully "get" my interest.

"Would you do me a favor, Bob?" I asked. "Would you call me after the sales meeting so that I know how many stations we sold?"

Bob's eyes were roving around the room. He said, "Sure."

Well, I knew that "sure," because I had heard that "sure" before. This guy wasn't going to call me—he was just being polite, thinking of a dozen other things as we stood there together.

What could I do to make sure this guy called me? Of course. If you're me, this is what you do. I smiled and said to him, "Bob, do you have a card?"

"Sure," he replied, taking out a card and handing it to me. He was still preoccupied, looking around, seeing where his salespeople were and with whom.

I jotted a little note on the back of the card, handed it to him, and said, "Hold on to this, Bob. All right? Talk to you later."

Bob said, "Sure," again, and slipped the card into his pocket as I walked away. I went back to the hotel. At about 7 o'clock the phone rang and it was Bob Raleigh. He was laughing.

I had written on the card, "Call Judy at 7 o'clock or your balls will fall off."

What I was really saying was, "This is who I am. I'm a funny gal, but don't screw around with me. I'm working my butt off just like your salesmen, so don't treat me as if I were some flighty talent that's here today, gone tomorrow."

I was able to make my point using humor that was disarming. It worked, and now Bob Raleigh and I are good friends. We're a team.

How to Get a Raise

Sit down, take a pad of paper and a pen, and make a checklist. How long have you had the job? What have you learned since being there? How much expertise do you bring to your job? Do you have a high level of credibility with clients of your company? Do you have a high level of credibility with the vendors who do business with your company? How much have you learned that's intuitive, that's not written down anywhere? How much does the company rely on you?

There are ways to judge that. For instance, if you say, "I'd like to take my vacation the last two weeks of July," and the response is stunned disbelief and panic, think about it. Make a list for yourself. How much responsibility do you have? How much are you relied on? If you're not sure, consider what happens on your days off. Does chaos reign?

There are business people who will tell you, "No one is irreplaceable." I always think that's an incredibly short-sighted point of view. I understand that replacing one "somebody" with another "somebody" is always easy, but replacing somebody who has your particular wealth of information, skills, and personality may be more difficult. And it doesn't make any difference what it is that you do.

Let's say that you clean houses for a living. That's your business. How hard is it for your boss to replace you? Many housekeepers think, "Well, I could be replaced in a second." But that's not true. You're indispensable once you know the routine of the house. You know what's valuable and what's not, what has to be given special care, what requires this polish and what requires that cleaning solution. You know where things go. You have earned the family's trust. I happen to know that many physicians have contemplated suicide after losing a receptionist. They tell me it's extremely hard to find medical office receptionists who possess the right combination of savvy, humor, compassion, and toughness. Thirty people may apply for a job as a receptionist in a medical office. However, out of those thirty, it's not easy to find the one intuitive person who can negotiate between such eccentricities of the physicians and the patients as the physician who's always late and who insists on overbooking, or the patient who always shows up fifteen minutes early demanding to be seen immediately. A doctor's receptionist needs to have the skills of a traffic controller and the compassion of a saint. It's a hard ticket to fill. What you might consider a menial job is not menial if you are, even in the short term, irreplaceable to your boss. Would there be a period of discomfort, an interruption in the flow, if you weren't there?

Decide what your bottom line is. Are you prepared to stick with it? What if the boss says no? Are you prepared to leave the job? Are you issuing an ultimatum, or just testing the waters to see what your boss thinks you're worth?

It's about you. Do you enjoy the work? Do you need the work? Would you be in sudden financial trouble if you didn't have the job? The rhythm of life usually dictates these kinds of moves. Think it through. The worst possible time to ask for a raise is in the heat of the moment, making your request in an accusatory tone—"I'm tired of bending over backwards and getting paid peanuts for all my effort."

Be crafty. Know the lay of the land. A woman who works for a publishing company once told me that she knows not to ask for money when her boss has his weekly financial meeting on Monday. On the other hand, when the sales figures come in on Wednesday, if the company has had a good week, he's downright benevolent.

FUN AND GAMES

The climate in many companies today has become somewhat grim because of the proliferation of sexual harassment suits. I'm not denigrating real episodes of sexual harassment, but so much of what is now being called sexual harassment isn't sexual harassment. In Yiddish, it's called *kibitzing*—kidding around. If you're walking the halls of your company with your sexual harassment radar ready to beep at any lapse of decorum, you aren't ready to play in the real world. Let's face it. The male sense of humor finds funny all that is basically scatological, meaning that the majority of male workplace humor is zoned in either bathroom jokes or sex jokes. There is occasional ogling and cracks to deal with as well. They're boys. You're a girl. Just pretend they haven't gotten out of junior high yet, and you'll begin to get the idea. You can't handle junior high school humor?

Don't get bent out of shape. Just practice that look of disdain you knew so well when you were twelve.

The Chronically Offended Woman

Speaking of sexual harassment, just because someone you work with does or says something offensive, it's not necessarily worthy of a lawsuit. A complaint maybe, but a lawsuit? Cases concerning improper behavior in the workplace come up on the *Judge Judy* program occasionally. I'm always surprised how little people understand what sexual harassment is, and what it is not.

Here's an example. A case came before me on *Judge*

Judy. A young woman was bringing a lawsuit against a male coworker. He had been attracted to her, had called her at home several times, and eventually asked her out. She accepted.

They went out on a date once, but she didn't particularly like him, so the next time he called, she said she was busy, and this continued for a while. She never told him that she wasn't interested. She explained to me in court that she was embarrassed, and she didn't want to hurt his feelings. Instead, when he sent her a humorous card in the mail, she sued him for sexual harassment. (Of course, being sued didn't hurt his feelings at all, right?)

I asked the young woman, "Why do you feel that you've been sexually harassed by this man? He asked you out, he sent you a funny card. I don't understand."

She batted her eyelashes at me and said with a quivering voice, "Well, it *felt* like he was harassing me."

There you have it—the absurd result of the national pastime of navel gazing. If you *feel* it, it must be so.

I frowned at her. "No. According to your statement, you *felt* embarrassed, you *felt* that you didn't want to hurt his feelings. You never said the three magic words, 'I'm not interested.'" I glanced at the unfortunate man she was suing.

"You, sir, may not be very adept at picking up subtle clues, or even not-so-subtle clues, that a woman is not interested. But that doesn't make you a sexual harasser. Case dismissed."

STAND TOE-TO-TOE

When I started in the television business, I was already a fully cooked human being. I thought I was pretty smart—at least as smart as the characters in California. They preferred to treat me as a rare and delicate flower, whose brain had just been sucked out of her head. They would constantly say, "Don't you worry. We're experts at this. Just rely on us."

I put up with it in the beginning, but once my show became a big hit, the condescending attitude started to get on my nerves. Finally, I said to my cadre of experts, "Let me tell you guys something. I'm a five-foot-one, fifty-ish lady who's not particularly gorgeous, but attractive. I come from a middle-class background. I wanted to be a lawyer; I became a lawyer. I wanted to be a judge; I became a judge. I wanted to run the courts in Manhattan; I ran the courts in Manhattan. I set my sights on having a television program; and now I have a television program. That's a hundred percent success rate.

"Now, you folks decide to produce, what—ten pilots for new shows every year? If one of those ten gets on the air, you're doing well. The odds of the one that gets on the air being a hit and getting renewed for a second season aren't good. So, although you're the experts, and this is your business, you have about a ten percent success rate. That means that ninety percent of the time, you fail. On the other hand, my show is a hit. So please don't treat me like an idiot, and don't try to turn me into someone else. This is what you bought, and this is what's successful."

It may have seemed a patronizing speech to give to a roomful of professionals, but I felt it necessary to make my point, and judging by the silence that followed, I did. Since then, I've been treated with less condescension and received more respect. Sometimes you just have to demand it.

It's important to know your worth. But more than that, it's important to let others know your worth.

One more story. The first week I was taping the show, a stunningly handsome young man was assigned to oversee my first photo session. Meticulously coiffed and wearing a silk suit and custom-made shoes, he was just beautifully turned out. He came up to me and said, "Judy, I want you to put yourself completely in my hands."

For some reason, this struck me as hilarious. "Honey," I told him, "I've got food in my refrigerator that's older than you."

The expression on his face was priceless. He fled. I

never saw him again. Wherever you are, I'm sorry if I hurt your feelings.

That's the way I am—what am I going to do? Change now?

GOOD OLD GIRLS

Based on my experience of the people I've met thus far in the entertainment industry, I'd estimate that women are far smarter and far more knowledgeable about the industry than their male counterparts. Women are also more likely to get the short end of the stick.

In the television industry, women are notoriously underpaid compared to men—from the top female talent on down to the female production assistants. What's the problem? Women tend to look for acceptance before compensation. A woman can be told that she's loved, and needed, and brilliant, and wonderful, and that will be enough to hold her down. Women are hypnotized by that kind of language. It feeds the innate need to satisfy, to give, and to nurture. Nurturing is fine at home, but you have to fight the impulse to bring it to work.

Since so many women aren't fully aware of their own worth, and lack that deep self-confidence that so many men are imbued with from the day they're born, they're prepared to accept less for the same work for which their male colleagues are being paid exceptionally well. Too many women in the business are still willing to settle for a stroke instead of a contract, a pat on the head instead of a substantial raise or bonus. Don't do it. Dogs and children get pats on the head. Women in business get cash.

BE PREPARED!

The secret to success is preparation. By that I mean structuring your life for all eventualities. Keeping your sensors alert.

Here's an example of what I mean. Let's say you aspire to be a bank teller. You like interacting with others, and you enjoy touching money. You get a job in a bank and you're happy. Then the bank sets up an automatic teller machine in the outside foyer. You don't pay much attention. It's just a convenience. Banks will always need tellers, won't they?

Time goes on. Two more automatic teller machines are installed. Your bank develops an on-line banking service. You're still standing at the counter, counting money and chatting with customers. You even hand them the bright little brochures with their bold offers: BANK ON-LINE! 24-HOUR AUTOMATED TELLERS! THE FUTURE OF BANKING IS YOUR HOME COMPUTER!

One day your branch goes fully automated, and you're out of a job. You can't believe it. How did this happen? Why didn't someone tell you? What are you going to do now?

Technology is developing so rapidly that none of us can afford to get set in our ways. Instead of hoping your world won't budge, stay alert to the new possibilities that are emerging. Don't get caught staring at a pink slip and wondering where you were when the rules changed.

It really isn't enough to be "certain" of what you want in life. It's the unexpected twists and turns, the things you least expected to happen, that very often turn out to be the most important—and sometimes even the best—events in your life. I was certain what I wanted in my life, and I got much more than I bargained for.

Nothing is sacred. I wouldn't be surprised to see electronic courtrooms in our future. That would mean a lot of black robes gathering dust in closets.

"SAY GOOD NIGHT, GRACIE"

Comedian George Burns died only recently. He was one hundred years old. In the fifties and early sixties, he and his wife, Gracie Allen, who was also his vaudeville and radio partner, had a long-running weekly television series called *The Burns & Allen Show.* George Burns always played the suave, acerbic, cigar-puffing straight man to his brilliant wife Gracie's silly-ditz act. Every program ended with a little series of non sequiturs that was eventually nipped in the bud when George turned to the camera and said, "Say good night, Gracie," and, of course, she did. They had a set routine that always ended their final bit. They knew when to get offstage. And, maybe more important, they knew how to get off the right way. They always left their audience wanting a little bit more. That's kind of how I felt when the opportunity arose to leave family court in New York City and transfer my skills as a judge to a television studio in Los Angeles. I really wasn't thinking about leaving the bench that year, but I knew the time was nearing when I would want to get out. My frustrations with the system, the bureaucracy, the court administration, had been mounting steadily for years. And the daily grind, the emotional toll, also began to wear me down. I'd been a good public servant for twenty-five years. I always gave one hundred and ten percent of my concentration and energy to everything I did. No one should stay in a demanding job that involves the very lives of other people if she finds herself burning out. What occurs in some professions—the law is one, medi-

cine is another—is a form of combat fatigue. At some
point, the clamor of war begins to seem normal, and
that's when you're in the most danger. I looked around
me and saw many people who had stayed too long. They
were shell-shocked—plodding along on automatic pilot.
Practically speaking, I'd rather have a doctor in his fifties
perform heart bypass surgery on me than a doctor in his
seventies or eighties. Why? I don't want to think about
someone's hands shaking with a scalpel poised over my
open chest. I don't want someone who might become
unduly fatigued, or forgetful. I don't want to hand my
fate over to someone who says, "I've done this procedure
so many times, I could do it in my sleep." My luck, he
might try.

You have to know when to say good night.

On the Other Hand . . .

There is always a new opportunity waiting in the wings. I
think one of the most destructive concepts we communicate to ourselves and to our children is that every person
has one career, one chance, one make-or-break opportunity. We ask our kids, when they are barely past puberty,
"What are you going to *be*?" We force them to choose a
"major" and stick with it for life. We set arbitrary finish

lines for success. Woe be the one who hasn't "made it" by age thirty. There is something ludicrous about setting the bar for making it at thirty when the average life expectancy is seventy-four.

The truth is that each of us has the capacity, if we so desire, to follow different paths at different points in our lives—to put an end to one pursuit and begin another. When you understand that, you'll see that there is no such thing as failure. If you go into sales and you're lousy at it, you can change course. People find themselves in the wrong game all the time. It doesn't mean they're stuck with it. Explore your options. What do you enjoy doing? What are you naturally good at? What gives you satisfaction? Once you've answered those questions, it becomes a matter of preparing yourself for what you want to do. Go to school at night, network with friends, do whatever it takes to realize your dreams. It's a tragedy to work at something you hate doing. It's corrosive— you're miserable, and usually so are the people who have to be around you.

I had a little chat with a woman recently, while standing in line for a movie. She'd just turned sixty—and had also just graduated from medical school.

"A dream fulfilled," the doctor told me. I believed her, too. She radiated energy and life.

Life can be an adventure or a chore. If you stay in a

profession or a job that doesn't challenge and excite you because you're accustomed to it, it's routine, you've learned to live with it—you're selling yourself short.

There's no cap on success. The jury stays out till you take your last breath.

CHAPTER 5

You're the Trunk of the Tree

The trunk of the tree is the foundation, the heart, the core. Who is the trunk of the tree in most households, in most companies? The woman. Women establish roots. Women breathe life into families, projects, and organizations. If it weren't for the steady, consistent, energizing presence of women, the social order would topple.

But who takes care of the woman? Who nourishes the roots of the tree?

Women are the nurturers, not the nurtured. The nature of a woman is to be giving, so the vital sap that keeps the tree alive is always circulating, always being used to produce what others need to thrive. Who's nurturing the woman; who's tending the tree?

You are the trunk of the tree. You need to say to yourself, "I have to nurture myself." And if others do it, too, that's a bonus, but don't expect it. Men may learn a nurturing technique, but it doesn't come naturally. This wisdom usually occurs much too late in women's lives, after decades of frustration, disappointment, and heartache.

When I was a child, my mother was the one who kept everything going. I loved my father totally, but I didn't realize how important my mother was to all of us until after she died at the young age of fifty-seven. Her contribution was so seamless, so effortless, that it was virtually invisible while she was alive. Soon after my mother was gone, I recognized how essential her role had been in keeping the family functioning. My father was always in the spotlight, but now I saw how much of my mother's efforts involved making him look good, caring, and socially adept.

It was my mother who kept track of all the anniversaries, birthdays, and special occasions. It was my mother who made all the holiday preparations, organized the parties, kept in touch with friends, bought and wrapped gifts, and basically took care of all the family's obligations. She kept it going. In retrospect, I don't believe I truly appreciated her efforts while she was alive, until the very end. I'm sorry I never had a chance to acknowledge her essential place. Though I'm sure she loved my father, I think my mother was frustrated with him, because he thought *he* was the trunk of the tree. And she happened to know differently. She was just too much of a lady ever to point it out to him. My mother worked with my father. He was a dentist, and my mother ran his office. She was a very elegant woman, although she wasn't very interested in fashion. She wore her comfortable white nurse's shoes around, with her white uniform. She never wore any makeup. She was always doing things for other people—it was her nature, I guess.

Once, when she must have been annoyed with my father about something, she went out and bought herself a big diamond engagement ring. That kind of extravagance was completely foreign to her. She'd had a simple gold wedding band that she wore all the years they were together. I remember her telling me, "I wanted it, and I got it." After she got it, she wore it very rarely. A couple of years later, when she got sick, she sold it. It wasn't the ring itself she cared about. It was what it represented—a rare moment when she decided to be self-nurturing.

That ring stood for something. Buying it was the most purely self-indulgent action my mother ever took. I was reminded of it only recently when I went shopping with my best friend Elaine.

THE ENGAGEMENT RING

Jerry didn't give me an engagement ring when we decided to get married. I admit to you privately that I was a bit disappointed. One day last year, I went with Elaine to the jeweler and bought a big diamond ring. It was a real beauty. When I came home, Jerry was sitting on the couch reading the paper. "What did you do today?" he asked.

I said, "I went shopping with Elaine."

Not looking up from his paper, Jerry asked, "Where did you go?"

I said, "I went to the jeweler."

That got his attention. He put down the newspaper and stared at my outstretched hand with the ring sparkling away on my finger. And he said these immortal words: "Good for you."

That's what I love about Jerry. He knows he's not adept at the intramarital nurturing touches. But he wants me to be nurtured, even if I have to do it for myself. Was I repeating my mother's life? I wondered. I really wanted the ring. I'd never had anything like it before. Was it an indulgence? Could I have spent the money more wisely elsewhere? Of course. Would it have been better off in a Keogh or an IRA—saving for the future, retirement, my grandchild's college education, whatever? Probably. However, I opt for this position: Spending your money on things that give you pleasure once in a while is food for the soul. Don't feel guilty. Consider it a long drink of water for the trunk of the tree.

GIVE YOURSELF A BREAK

The most difficult time in a marriage is when there are young kids. A lot of it has to do with equity and responsibility. In the majority of marriages, women are still the primary caretakers of children, and as such they are the ones who make the greatest personal sacrifices. Despite the great strides women have made in the workplace and the world, the family scale of responsibility usually tips wildly out of balance. Moms do the cooking, the cleaning, the laundry, the child care. Dads are a lot better about "helping" than they used to be, but that's usually all it amounts to—*helping*. This fundamental inequity creates hostility. This is especially true if you're a professional woman who may take three to six months of maternity leave after the baby is born. Or you may quit your job altogether when the kids are young. Your husband goes off to his job, and there you are. It can be jarring. You're used to an active environment, an exchange of ideas, and suddenly you're spooning Pablum. You start to brood. You're confined. You have a little baby completely dependent on you, you have no private time, your conversations consist of "ooh-ga" and "ma-ma," and your career is on indefinite hold. You begin to resent the freedom that your mate has, and that you've been forced to give up in your new role as a mother. And to top it off, without your former salary, you're doing all this on bare bones. Yes, you may have chosen this, but it's impossible to fully prepare for it.

As the children grow older, the responsibilities and demands increase. There is more stress than ever in the relationship. The typical MO is that one parent plays the good cop—"I'll protect you from your mother's psychotic rages"—and the other becomes the bad cop—"You want to cry? I'll give you something to cry about!" Usually the father gets to play the good cop, because he's not involved in the day-to-day, minute-to-minute shenanigans. The friction between the couple develops when the bad cop tries to impose discipline, and the good cop steps in and protects the "criminal" child. It's like that climactic show-down scene in *High Noon*. My husband was always Gary Cooper, and I was always the bad guy everybody had been waiting for, getting off the train.

You spend your days in this atmosphere, giving fully of yourself and receiving little appreciation in return. The coup de grâce is when your partner comes home and asks you what you've been doing all day while he was working—in a tone that implies he's thinking, "Doesn't look like you do much."

My advice: Get sick. Develop a malady for a couple of days—preferably on the weekend. You're so weak you just can't get out of bed. Let Dad do it all. It's a real eye-opener for him, and a well-deserved break for you.

"You Go, Girl"

When I was approached about having a television show, I began to think about the message I wanted to get out to the world. When you've got such a big stage, you might as well make it count for something.

Still, I wasn't quite prepared for the level of enthusiasm I receive from young women. Everywhere I go, they cheer me on, saying things like "You go for it, Judge, you tell them!" Older women tell me that their daughters race home from school to watch the show. My sense is that I've become a positive role model for young women and young girls. That thrills me, because my message is a clear and important one: Women have the power to make decisions, to call it as they see it, to take no guff. I represent that to millions of young women and girls, and I'm proud. (I have also figured out why *Judge Judy* is popular with so many men. They watch me and they think, "Maybe what I have at home ain't so bad.")

I tell women to love themselves, to respect themselves, and to demand respect from others. Who's the trunk of the tree?

I have to admit, though, that it took me a long time to figure out this trunk-of-the-tree business. I really hadn't learned to take care of myself without feeling little pangs of guilt. In 1996, after I retired from the family court with its grueling ten-hour days and started doing the television program, I suddenly had a whole lot more time to myself than I'd been able to enjoy in well over thirty years, when a busy career and family life often tested my mettle. Initially, I felt almost guilty about it. I'd try to act busy, just for show. I've gotten over it, though. Now when Jerry comes home and asks, "What did you do today?" I might reply, "I went for a facial, then I took a little nap. Then I left the house, went to the dry cleaner, had lunch with Elaine, and got a manicure. Then I came home, read for a little while, and took another nap. That was my day." When Jerry looks around, comes back into the living room, and asks me, "Where's my dinner?" I usually say, "Across the street," meaning the pizza and takeout joint across from our apartment building in the city.

No excuses. No guilt. When I work, I work hard. But

in the past, I had viewed relaxing as synonymous with lazy, and now I know that's *crazy*.

LOOK GOOD TO FEEL GOOD

Let's talk about this beauty thing for a moment. Forget all the perverse and arbitrary expectations for women about weight and hair and skin and fashion. There is only one reason to look your best, and that is to *feel* your best. We are what we *feel*.

I take care of myself and how I look because it adds a bounce to my step. I like coming across to the world as confident. We all check one another out. You can look at someone and think, "This person dressed with some thought. She cares about herself."

Choosing the right kind of clothing for your figure, being able to apply makeup artfully, looking ready to meet the world—all these make a clear, if subliminal, impression on everyone you meet.

Even so, decisions that you make about your hairstyle and fashion choices shouldn't be dictated by the likes and dislikes of a partner. It's not really important that you look your best and be the best you can possibly be physically for other people. What's vital is that you do all this for

yourself. Even when you know you'll be completely alone
for a day or more, there's still no reason that you can't get
out of bed in the morning, wash yourself, put on a little
makeup, comb your hair, and put on something clean,
comfortable, and nice-looking. It's what that does for you,
not for others. It makes you feel good—awake, alert,
refreshed, crisp, and ready to take on the day.

RAPUNZEL

While I'm on the subject, I'd like to say a word about hair.
Lots of men just love long hair on a woman—they think
it makes her sexier, I suppose. If you happen to like the
way you look in a shorter, more manageable length of
hair, and the man you're with is crazy about long hair, buy
him a copy of the fairy tale Rapunzel. It features a beautiful young woman with enough hair to clothe the nation.
Do what works for you, not for him. If you were to mention that you thought Michael Jordan was fabulous-look-
ing, and you'd like to see what your partner looked like
with his head shaved bald—very *au courant*, by the way—
would he rush into the bathroom and make it so? Not in a
million years.

ENHANCEMENTS

According to the American Society of Plastic and Reconstructive Surgeons, the most popular procedure today is breast enlargement. Now, I don't have anything against plastic surgery. If a woman wants to improve the way she looks *in her own eyes*, I say go for it. But I have to wonder about breast enlargement, because big breasts are a male fantasy, not a female fantasy.

Last year, I had a case on *Judge Judy*. A woman was suing her doctor because, she claimed, after two breast enhancement surgeries, her breasts were not perfect. She felt she needed a third procedure to fix the problem, and her husband was standing by her side in full agreement. Since there were no photos, I asked the woman to step into my chambers so I could evaluate the evidence. Suffice it to say, these were the breasts of a goddess. If they were *uneven,* as she claimed, it certainly wasn't visible to the naked eye.

I returned to the bench, and I said, "I'm jealous. At my age, *even, uneven,* I'm just happy I have them." This statement was true, but there had to be more to her fixation. I focused on her husband.

"Sir," I said, "you encouraged your wife to see another plastic surgeon, to have a third operation." I was dismayed, and it showed. "You should have encouraged her to see a psychiatrist."

It was clear to me that the woman's obsession about her breasts was directly related to her husband's eager compliance. Instead of telling her she was beautiful, and that he didn't want her to get cut again, he was feeding her insecurity. He was the judge and jury of her slightest imperfection.

Ladies, do it for yourself, not for him. And remember, those *Playboy* pictures are airbrushed.

KEEP YOUR TEETH IN

I spoke with a woman who's been married for thirty-five years. Her husband has never seen her without full makeup on. She sets the alarm for 4:30 in the morning— she's completely done every day before her husband opens his eyes. Her husband rises and walks around *au naturel*. It's like Beauty and the Beast.

I wouldn't go so far as to get up at 4:30 A.M. to put on makeup, but I applaud the notion of always putting your best face forward. I've always tried to have my husband see me only when I'm at my best. I do that for a very good reason. When things are good in a relationship, your mate will always picture you in the most positive light. When things aren't going so well, there's a tendency to rehash all the bad times, revisit them, brood over them. Five years, ten years. We're able to recall in vivid detail days, times, conversations, gestures, looks from twenty years in the past.

Just as women remember every rotten, nasty thing their husbands have ever said or done, I have a feeling that, for men as well, such behavior is human nature. When things are rocky in a relationship, they search the deep recesses of memory to find something especially negative and ugly about their now unloved partner. If and when that happens, I don't want my husband lighting on the memory of a grotesque creature resembling me falling out of bed in the morning unwashed and uncombed. Why provide him with the free ammunition?

So I get up before my husband every morning and brush my hair and put on a little lipstick, so his first bleary view of me is pleasing. It makes *me* feel good.

I learned at an early age that a physical flaw can stick in your craw. When I was fifteen years old, I had my first serious boyfriend, Marvin. He wasn't exactly a winner, but you couldn't tell me that. I was fifteen and in love. My father knew Marvin was marginal at best. We were at the dinner table one night when my "romance" with Marvin was at its height. My father looked at me and said, "Do you think he spits through that hole in his front two teeth?" We were a very tooth-conscious family, my father being a dentist. Well, that was it. Every time I looked at

Marvin, I could see that his two front teeth were separated, and I couldn't help picturing him spitting through that hole. The romance was soon over. The thought stayed with me, and it affected the way I wanted to be seen by the man I love. So I brush my hair and put the lipstick on. My husband will never have an opportunity to see me spitting through the hole between my teeth.

LOOK TO YOURSELF

If the tree turns brown and dies, does it blame the branches? Of course not, but you'd never know it from listening to the machinations I hear in court. My experience on the bench could be summed up as "one thousand and one excuses." Excuses are the bread and butter of court life:

"I robbed a store because my mother didn't love me."

"I had a baby out of wedlock because I'm disadvantaged."

"I let him abuse me because I was trapped."

"I can't find a job because of discrimination."

I know that this view is controversial, but I believe victims are self-made. They aren't born. They aren't created by circumstances. There are many, many poor, disadvantaged people who had terrible parents and suffered great hardships who do just fine. Some even rise to the level of greatness.

You are responsible for nurturing your roots, for blooming. No one can take that away from you. If you decide to be a victim, the destruction of your life will be by your own hand.

CHAPTER 6

You Can't Teach the Bull to Dance

Once you understand that you're the trunk of the tree, you also have to face a terrible truth: Trying to change a man—to make him more helpful, more responsive, more socially acceptable, more sensitive, more domesticated—is about as feasible as trying to teach a bull the two-step. The result is going to be a pile of broken china and a load of irritation for *you*. I was enlightened about this fact during a rocky point in my marriage. My husband, Jerry, and I had reached a marital impasse, and after much cajoling, my "bull" reluctantly agreed to a session with a marriage counselor. We sat inches apart on a couch while I spewed out my complaints. Bottom line, he didn't understand me or my needs. Sound familiar? My handsome, adorable bull grunted often, was visibly uncomfortable, but was captive for a full fifty minutes. Jerry and I had been married for fifteen exciting, interesting years, but for all that time I indulged in the female struggle to make him think like a woman. The session was almost over when the therapist reached for a large bowl of grapes. He handed it to my mate and instructed him to slowly feed me one grape at a time, and I was to accept his offering without touching or helping him. The symbolic nature of this exercise did not escape me. He was giving, I was receiving. Not wanting to insult the affable therapist who looked as if he had just reinvented the wheel, we concluded the exercise, thanked him, plunked down one hundred bucks, and left. For the next week, every time I started to complain about his lack of understanding, empathy, caring, Jerry would whip out a small box of raisins (he had decided grapes were too messy) and demand that I sit down for a feeding. But the exercise didn't inspire the promised intimacy. It was just plain irritating. And then it hit me—this was the turning point. I was fifty years old. I finally concluded that the struggle was over. It struck me like a bolt of lightning. I'd spent most of my adult years trying to teach my chosen bull to dance. Whether owing to nature or nurture, I just couldn't get this bull to do the cha-cha. So if I wanted inner peace and happiness, it had to come from me. Why had I ever expected him to provide it for me? He was just the way he was, and no matter what I did, he wasn't going to change all that much. I thought about the thousands of troubled couples I had seen as a family court judge, and all my female friends whose basic complaint always boiled down to the same thing. You can't teach the bull to dance. We must raise our daughters to get on with their lives—and not be stalled by the same bevy of frustrations that have paralyzed women for generations. Women are still looking to men as the source of all meaningful approval, as the beacons of light in the deep, dark cold of outer space. It's just not so. Love has to emanate first from within yourself—we have to teach our young women to love themselves and respect themselves. Confirmation of your value as a human being doesn't depend on the approval of any man, be it your husband, brother, father, or boyfriend. Women are complete individuals without the need of men to establish their purpose and direction in our society. Times have changed. The relationships between men and women are changing as well. The old expectations, the old contracts, no longer necessarily apply. Most of us want our mates to complement us, to make us feel as though our lives are balanced and complete. If you expect your man to understand your every subtle emotional turn, and always to treat you as a completely equal partner; if you expect him to empathize when you're having a particularly nasty bout of PMS, cramps, hot flashes, menopause, or the thousand other little hells only women are prone to, you will end up feeling frustrated, disappointed, and unhappy.

I deferred to men for years because it was expected, it was expedient, and it was necessary to keep the peace, and I resented it. It was expected of my generation. Feminism and women's liberation are only words out there in the real, everyday world. Ideals don't always play out as well as they sound. Inner peace truly comes when you recognize and accept the differences between men and women, and decide to enjoy the filet and discard the gristle.

Here's some food for thought. It has been said that there are two kinds of men: those who don't get it, and those who *do*, but get it wrong.

LIFE'S *SCHMUTZ*

Schmutz refers to the dust balls that keep building up in the corners of your life, the petty annoyances. All the little stuff, the debris caused by trying to accommodate everyone in your life. *Schmutz* is usually women's work. We do it for two reasons: one, so it will get done; and two, to keep the peace. Men are never going to take on the *schmutz*. It's not their job. You have to come to terms with the fact that in relationships there is almost never an equal division of responsibilities. It's not going to happen. It probably never has and it never will. Eventually, you've just got to make your peace with it. Look at all the good things you get in a relationship. If you're lucky, your mate gives you a good laugh. Also, if you're lucky, he cleans up nicely and you can take him to the occasional social function. Very important. If he keeps your feet warm on cold nights, that's also a bonus. And if he cares about you and you him, if you rely on him and he relies on you, then you're doing really well. Just reject the notion that love and marriage is a fifty-fifty deal—it's not. If you fight it, the only one who's going to suffer is you.

Figure out a percentage you can live with. If you do sixty percent, and he does forty or even thirty-five percent, and that works for you, fine. But if you're doing seventy-five percent, and he's barely amping it up to twenty-five percent, that may be different. The real question is, what gives you peace of mind?

A long time ago, I decided it was easier for me to do the vacuuming myself rather than ask Jerry to do it. If you've ever seen a man running a vacuum cleaner back and forth across one small area of carpet, while his eyes are riveted to the television, you know what I mean. It also became easier for me to clean the bathroom myself rather than finding Jerry in there wiping a spot on the mirror with the kitchen sponge, which he had just used to clean the toilet.

"That's the kitchen sponge."

"Huh?"

"That's the kitchen sponge. The sponge you just washed the toilet seat with, you got that in the kitchen, didn't you?"

"Yeah. I did. I guess it is the kitchen sponge. Now it's the bathroom sponge."

You get my drift. The level of annoyance isn't worth it. Curiously, my children tell me that when Jerry and I were divorced, he became Mr. Clean. He'd greet them at the door of his apartment with a bottle of Windex in one hand and a cloth in the other. He'd follow them around, wiping every smudge and admonishing them not to touch. What happened to *that* guy? I never found out. I guess he got lost somewhere between Fifth Avenue and Park.

Sometimes it's easier to do it yourself. However, there are exceptions.

Don't Get Trampled in the Process

Just because you can't teach the bull to dance doesn't mean you have to let him run wild and trample the petunias. In the years I have been married to Jerry, there have been occasions when I have had to reconfigure the percentages on our domestic landscape. This has usually occurred when an inconsequential, mundane matter started feeling like an albatross around my neck.

For example: the shirts. Until a few years ago, I used to handle all the dry cleaning and laundry. Typical wife stuff—although I will add that we had the same position, both judges, and the same salary. I would send his shirts out to be done along with my blouses, suits, and dresses.

He was never satisfied with the result: a button missing, too much starch, not enough starch, they pressed the collar wrong. Every week we'd get the dry cleaning and the shirts; every week we'd spend five minutes on each shirt as Jerry pointed out all the cleaner's flaws. We'd have long, complex discussions about the various pressing techniques, an errant wrinkle, until he'd say, "I want you to bring this one in and show him this. And bring this one back, too." And for years, I did this. I'd change dry cleaners, looking for a place that would do Jerry's shirts as he liked them. Then, inexplicably, I said to myself one day, "Hey, Judy, what are you? Some kind of an idiot? What are you aggravating yourself over this for? You don't have enough to do? You need to find the perfect dry cleaner, too?"

Call me passive-aggressive. I didn't say anything to Jerry. I just stopped taking his shirts to be done. I let them pile up on the bottom of his closet floor for a few weeks, until he was putting his neckties on over T-shirts. Then it began to dawn on him. "Honey?" I heard him yell inquisitively from his closet one morning, "I don't have any clean shirts. What happened to my shirts?"

Eventually, he came into the kitchen, where he found me making coffee.

"Honey?" he began again, "All my shirts are dirty."

"I know that," I responded.

"What am I going to wear to court today?" Jerry asked plaintively.

"Why don't you go through the pile on your closet floor and see if you can find one that fits the bill?" I offered helpfully.

"All my shirts are dirty," Jerry repeated to me wonderingly.

"Your shirts are dirty because I haven't been bringing them to the dry cleaner. Ask me why," I said to Jerry.

He took the bait. "Why?"

"Because I've gone to twelve different dry cleaners in

the last three years, and none of them can do your shirts the way you like them. I'm persona non grata at all of them. So I'm not taking your shirts to be laundered anymore. I'm not aggravating myself. It's annoying when I have to go into the dry cleaner and deliver your litany of complaints. They're not happy, and I'm not happy. I get aggravated when I go and pick everything up, because I know I'm going to end up getting an earful from you. You know, there are fifteen dry cleaners in the immediate area, and I've been to almost every one of them. As a matter of fact, I have one now that I like. I'm not going to tell you where it is, because if I bring your shirts there, it'll be all over in a month. I can't keep doing this—I have to simplify my life."

Now Jerry takes his shirts to the dry cleaner himself. A different one from the one that I patronize. I still won't tell him where I go. Believe me, it's better this way. I'd be willing to bet that he never utters a word of complaint now that he has to face them himself. I wouldn't go so far as to say that he's changed, but at least I've eliminated one irritant from my life. And that's the point. Sometimes it's easier to do a task yourself when it will drive you crazy waiting for him to do it right. But when it's his "stuff," you can happily let it go. It felt great to eliminate the weekly shirt crisis from my life.

THE *TITANIC*

Likewise, the boat. I'm a big fan of canoes. Low mainte-
nance, quiet, portable, no moving parts, you don't have to
clean them. Grab the oars, get in, paddle away. A self-
propelled activity that requires balance, symmetry, and
muscle.

We bought a dream house by a lake in upstate New
York. It came with a small canoe, which Jerry and I pad-
dled through the still waters on warm summer evenings. It
was my idea of heaven. But the family—especially the
men—wanted a *real* boat, the kind powered by fuel and an
engine. I resisted, but was eventually talked into it by the
concerted efforts of my husband, kids, and grandkids. "We
can go water-skiing! The babies can float in tubes tied to
the boat. We'll have so much fun!" So we bought a boat.
Less than a week after we got it, Jerry destroyed the pro-
peller and the shaft on some rocks that materialized out of
nowhere. The boat needed extensive repairs. How is this
accomplished? Not so easily. It's a big boat. It needed to be
brought out of the water, placed on a trailer, transported
to a boatyard for repairs, then, once repaired, placed on a
trailer, transported back, and relaunched into the lake. Of
course, all this was arranged and handled by yours truly.
Ten days later, Jerry and the kids came in after an afternoon on the lake. Jerry said, "You've got to call the boat-

yard again. The engines aren't right—they're skipping. Tell them to come and get it, pull it out of the water, and repair it."

I'd gotten a lot of sun that day. My brain was hot. I exploded. "As far as I'm concerned, you can plant geraniums in the damn boat. I'm not calling, I'm not waiting around for the boatyard guys, I'm not having long discussions with the mechanic at the boatyard about inboards versus outboards. I'm not calling 1-800 numbers and waiting for hours to order a manual for the boat. Nothing! You want to play with this boat, you want it so much, you make the calls and arrangements, you be here for the pickup and delivery and the rest of it. I'm done. I'm divorcing the boat."

Ah, what a load off that was. The sense of freedom I felt may never again be duplicated in my life.

Here's the lesson. We encourage learned helplessness in our husbands and children because it makes us feel needed. For some reason, we think that if we take on all the responsibilities that we can, our husbands and children will adore us that much more. It's a total falsehood. As a wise friend once said, "Men don't run home to clean a bathroom or to make a hot meal." Sooner or later, we will resent taking on so much for so little reward. Have you ever noticed that it's the doers—the active partners, the ones who take on the most responsibility—who come in for the brunt of the criticism? Turn it around: "If you don't like the way I'm doing it, then do it yourself."

THE FEMALE CONDITION: HITTING THE PORCELAIN

There are certain truths about men that are eternal. Like the tendency to leave the toilet seat up. Women have debated for decades about the complex psychological basis for this flaw. We have also tried nagging. All to no avail. The position of the toilet seat remains up.

I, too, have fought this losing battle. And while I acknowledged that I could not change Jerry's behavior, I could certainly make him think about it twice.

One night years ago, after Jerry and I had divorced and remarried, and long after I'd recognized the importance of loving yourself, I had to use the bathroom. It was midnight, and we'd gone to bed a couple of hours before, but I knew my way around in the dark. I found my way to the bathroom with my eyes still shut, sat down, and practically fell into the toilet. Now I was awake! Jerry had left the seat up. Again. I could almost predict with uncanny accuracy when Jerry might pull this on me. It was his personal passive-aggressive statement. He'd be annoyed or angry with me about something, and instead of arguing with me about it, he'd just leave the toilet seat up. That particular night was a sneak attack. I hadn't expected it, so I was unprepared for hitting the porcelain. What woman living with a man hasn't experienced this? We've all felt the cold sting of porcelain in the dark of night, the lurching sensation of having fallen through a hole in the ice. To say that it annoyed me is an understatement of vast proportions. I hauled myself up and put the seat down. After I had taken care of my personal business, I went into the kitchen, turned on the light, and started searching in the cupboards. I chose two fairly large pot lids, then quietly returned to the bedroom. Jerry was sound asleep, sleeping the sleep of the good, the innocent, the pure. I quietly got into the bed. Then I turned over and straddled him, banging the pot covers together like the cymbals and cannon fire at the climactic finish of Tchaikovsky's *1812 Overture*.

Even with my weight on top of him, Jerry practically jumped out of the bed, his eyes as big as manhole covers. I was delighted. He stumbled out of bed and stood there looking at me as if I were crazy. By this time, I was standing on the bed, pot lids still ready if he didn't get my message loud and clear. "What are you doing?" Jerry screamed at me, giving you the polite version of what he actually said.

"The next time you think about leaving the toilet seat up," I said, brandishing the pot covers, "I want you to remember this moment. Your heart's pounding, isn't it? I woke you up, didn't I?"

"Yeah, okay," Jerry muttered, as his eyes began to turn sleepily toward the bed. I crashed the pot covers together three times more. Jerry looked wide awake.

"No, not okay," I threatened, standing on the bed. "If I fall into the toilet, it wakes *you* up, too."

It made an impression. I have to tell you truthfully that my midnight symphony was not one hundred percent successful, but things got better. I still had to be alert. What did change things was moving to a place where we have separate bathrooms. I never set foot in his. He never sets foot in mine.

If you're not lucky enough to have separate bathrooms, you might take heart from this news. Recently, I heard about a new invention soon to be marketed by a man who has a wife, daughter, and many sisters. He proudly calls it the Considerate Toilet. By some mechanical genius, the toilet senses when the lid has been left up, and after four minutes it automatically closes. Men are good with tools to mask inconsideration. Still, it takes care of the problem.

IN AND OUT

I sometimes joke that within the family structure, what men do best is breathe. They inhale and they exhale.

Male bashing? No. Truth. Women do everything. Work, shop, cook, clean, do laundry, take care of the children, arrange social functions, the whole *megillah*. Men breathe.

If there is ever to be a solution to this imbalance, it will have to start very early in a boy's life. Mothers have to teach their sons to become dual-process thinkers, because by the time they're adults it's way too late. Example: A man is standing in the bathroom relieving himself, and he reaches down to grab a piece of toilet paper, only to find the roll almost empty. Here's where the dual-process kicks into action. First, the male recognizes that there is a deficiency of toilet paper. Second, *and this is vitally important*, his next thought is to replace the roll of toilet paper, and *he does!* Which brings me to the subject of the toilet paper wars.

THE TOILET PAPER WARS

Question: How many men does it take to change a roll of toilet paper?

Answer: No one knows. It's never been attempted.

Maybe while he's at it, the inventor of the Considerate

Toilet can design a Considerate Toilet Paper Changer. Toilet paper often becomes an issue of great significance in relationships. Women need a good amount of it. Men can get by on one roll for months, sometimes years. Jerry would always leave the roll with one torn, half-glued sheet sticking to the bare cardboard beneath, sitting pathetically on the holder. If I said anything, he'd protest, "There's still paper on the roll." Technically, he was correct. Finally, I decided to test Jerry's definition of "paper on the roll." I left the roll with the one half-glued sheet there, and kept my own supply of toilet paper hidden. I saw him surreptitiously trying to find the toilet paper a number of times, without success. I patiently waited.

"Honey?" I heard Jerry calling from the bathroom.

"Yes?" I sweetly responded from the other room

"I need some toilet paper in here," Jerry called.

At last.

"There's still paper on the roll, honey," I called back with a big smile on my face.

Revenge is sweet.

ROAD RAGE

Have you ever noticed that the kindest, most gentle men become rogue warriors when they are behind the wheel of a car? Of course you have! We've even invented a name for the malady—road rage.

While women sit white-knuckled in the passenger seats, with their eyes squeezed shut and their feet pressing down on imaginary brakes, their mates play a high-speed game of bumper cars. And woe be it to any driver who tries to pass. It's never a pleasant ride. The more we beg men to slow down, the more belligerent they become. As our lives pass before our eyes, we remain mute. The awful truth of the matter is that even when a woman's life is in the balance, she still wants to be liked.

I practiced the silent death ride for many years before I finally decided that I'd had enough. While traveling on a narrow, winding, two-lane highway, Jerry decided to play kamikaze hopscotch with a red Porsche. I was terrified, but I managed to summon up the voice to say, in a matter-of-fact tone, "You know, dear, if you pass the Porsche, you're still going to be short."

That was it. I had finally told the truth. Jerry laughed, and he behaved himself for a good ten miles.

THE BULL'S WAY

One more story. And, by the way, if you think I'm too hard on my beleaguered mate, rest assured that no one finds these stories more amusing than Jerry himself. Maybe he can't clean a toilet bowl, but the man can laugh, which, in my opinion, is a very good trade-off. One morning in the early days of our marriage, as we were driving to work, Jerry idly mentioned that I'd mismatched a pair of his socks the last time I'd done the laundry. In fairness, he didn't say it with rancor or even disapproval—just a vague wonderment, as if to say, "How do these things happen?"

By this point in my life, I was quite practiced in reading between the lines, and I reacted accordingly. After letting loose with a tirade that included words like "spoiled," "selfish," and "thoughtless," I ended with, "You aren't rich enough to act like such a prince. Do your own laundry from now on." He shrugged with that maddening air of indifference. "No problem. I will."

A line had been drawn in the sand, and he had walked over it. In the weeks to come, I paid no attention to the state of my husband's laundry. I made it a point to forget that it ever existed. As far as I knew, everything was fine. But one day, weeks later, on a rainy Friday afternoon, I picked Jerry up from work. Before he even got into the car, I could tell that he'd had a terrible day. He looked pale and drained, miserable, as though he were in pain. "What's wrong?" I asked, figuring I could console him by letting him pour out his day's difficulties to me. But he was unusually intractable. He merely grunted a monosyllabic disclaimer and turned away silent, lost in thought. Nothing further was said during our drive from lower Manhattan uptown to Riverdale. We parked, took our briefcases and coats, and went to our apartment. As usual, we rushed into the bedroom to take off our business clothes and change into our jeans. I took off my suit jacket, my blouse, and then my skirt. Standing in my slip, I reached beneath to pull off my pantyhose when from behind me I heard Jerry say, in almost a whisper, "Do you want to know why I'm in such a terrible mood?"

I turned around and looked at him. I'd never seen him look quite so miserable. I sat down on the bed, looked up at him, and gave him my best sympathetic look. "Around noon today," he began, holding his tie in his hands, "the damn zipper on my pants broke. No big deal. I buttoned my suit jacket over it. At lunch, I went to my law secretary, and I asked if he had a needle and thread I could borrow. I figured I could repair the problem with a quick sewing job."

Apparently, the law secretary did indeed have a needle and thread, and he gave them to Jerry. Jerry then went into the men's room to do the repair. He took off his pants so he could do the sewing, and was standing there in a shirt, tie, and suit jacket, busy with the needle and thread. Suddenly, the door to the men's room swung open, and two burly Secret Service agents came in. They had an office on the same floor of the federal building as Jerry did.

"I was never so mortified in all of my life," Jerry said.

"Jerry, Jerry," I quickly replied sympathetically, "don't be ridiculous. I'm sure they understood."

Jerry locked his eyes on mine. Then he said, "You think they understood *this*?" And then he dropped his pants. My gorgeous husband of six months was wearing a pair of my lacy pink bikini underwear. And I'm a size two.

So the truth came out. He'd never done his laundry. As I said, you can't teach the bull to dance. When he ran out of underwear, he started wearing mine—until that Friday. For the next two years, Jerry took the elevator two floors above his own in the federal building, just to avoid ever having to run into those Secret Service agents again.

ADDENDUM

Jerry and I have been together for twenty-two years now. He's a judge in the Supreme Court in the Bronx, and until I began appearing on television in *Judge Judy*, I was the supervising judge of the family court in Manhattan. We have exciting, interesting careers, each other, five wonderful children, and, so far, four adorable grandchildren. I still do the shopping. I still clean the house and act as the social secretary. I'm happy to report that I did give up cooking. And I'm pleased to report that my husband has completely mastered the Zen art of breathing. Life is good.

CHAPTER 7

Failure Doesn't Build Character

When I was four years old, I was enrolled at Miss Nod-didge's Dancing School, right across the street from where I lived, at 225 Parkside Avenue. Dancing schools were very popular when I was growing up. Every little girl in the neighborhood was sent to dancing school. Unfortunately, I had legs like a piano stool and wasn't terribly well coordinated. Apparently, I was never destined to be a prima ballerina.

I participated in all the exercises and classes. I studied ballet, tap, and acrobatics—and I didn't exactly fly through the air with the greatest of ease. After one fairly severe injury—caused by a failed double back flip—I was excused from dancing classes until I could bring a note from the doctor. My parents were smart enough to realize that maybe dance and acrobatics weren't for me. So I was allowed gracefully to withdraw—until then the most graceful thing I'd ever done—from Miss Noddidge's Dancing School. Until I could walk again, anyway.

My parents continued to provide me with a smattering of lessons—piano, violin, swimming, and so forth. It took them a while to recognize and fully value my real talents. From my experience as a child, I firmly came to believe that, from the beginning, you must find the few things a particular child is naturally good at, and cultivate those things. By doing so, by developing a child's natural skills, you provide your children with the confidence to take healthy risks, to push the boundaries, to learn on their own.

Short, stubby little girls with a lack of coordination don't often blossom into ballerinas. It's not exactly a confidence booster for them to be surrounded by lithe, graceful bodies executing perfect pliés and twirls, while they're stumbling around looking like the Pillsbury Doughboy in toe shoes. Over time, my parents figured it out. My father came to realize that I had a gift for languages. I also had strong communication skills, but I showed no flair for math or science. I got through those subjects, with some difficulty, in high school and college. I had tutors, I had special help after school, and I took extra remedial classes. My parents were always supportive, because they knew I was trying. They never once said to me, "How could you get this C in chemistry or that C-minus in math?" I excelled in debate, social studies, and history, and I took my intellectual encouragement from those venues. My parents praised me for my work on those subjects. They knew how hard I struggled with my math and science courses, but understood that my talents lay in other directions, as evidenced by my interests and aptitudes. They accepted that I wasn't particularly gifted in some subjects. Although they would have preferred it if I'd done better in them, it wasn't that important. As a result, I was able to view myself as a successful orator instead of a mediocre mathematician, an engaging communicator instead of a failed ballerina.

My Built-In Cheering Section

Being happy doesn't take money, although it helps. It doesn't take fame. Happiness requires an innate sense of security. Security comes from liking yourself, believing in your worth as a person. Security was the gift my parents gave me as a child—and that included the gift of feeling special.

My parents were tremendous supporters and boosters—they gave a lot of love combined with intelligence, compassion, and wisdom. I was the older of two children. We were Brooklyn middle-class. My father was a dentist, and my mother was a homemaker—and what a homemaker! My mother was the winner of the family show—she did her part and more! She was the one who always brought all our extended family together. She'd plan and cook for holidays, made sure everyone was well fed and taken care of, and she still always had plenty of time to shower her children with love and attention. My father was an intelligent, good-humored man who always doted on me. He wanted to make me feel special, and he did. I thought my parents were fabulous, and they gave me a good, solid grounding in life. They always did their best for me. In adulthood, I've always been able to point to their boundless nurturing as the source of my "can-do" spirit.

As a judge in the family court, it often amazed me that some children managed to survive the worst parents. I would see an abusive, drug-addicted mother with a teenage son who was compelled to run the household. He'd take care of younger siblings, nurse his mother, and fulfill all those roles while also excelling in school and in the community. Sometimes the worst parents had the best children. These children were much like the rare and hardy flower that struggles to push through a narrow crack in the concrete and bloom. They weren't the norm. If the parents were dysfunctional, the children were usually dysfunctional as well. I often had parents come before me in court, complaining, "I can't do a thing with this child."

My response: "Have you tried saying, 'You're terrific'?"

Their faces would take on puzzled expressions. They couldn't grasp the notion that a difficult child could be terrific. They were ready to praise the child once he proved his worth—not before. It was a sad catch-22, because in order to have the confidence to excel, a child must first be instilled with the knowledge that he has what it takes.

It broke my heart to see the cycle of defeat passed on from generation to generation. The combination of immaturity, lack of education and socialization, and absence of self-worth created an environment of hopelessness. If you don't have self-esteem, how are you going to instill it in the fragile psyche of a child?

QUEEN FOR LIFE

Barbara Corcoran is a dynamic, talented woman who, at forty-nine, is president of one of the largest real estate offices in New York City. Barbara has worked very hard to get where she is today, but she ultimately credits her success to the way her mother made her feel when she was a young child.

Barbara was dyslexic, but instead of bemoaning her daughter's plight and telling her what she *couldn't* do, Barbara's mother highlighted her special gift. "She always said that I had a wonderful imagination," Barbara told me, "and with imagination a person could do anything. I always felt that I had this tremendous asset."

Barbara's mother instilled in her the confidence and the courage to start her own business at the age of twenty-three. She never doubted that she would succeed.

My dear friend Vicki Schneps once told me how every year on the Jewish holiday Purim, her mother would tell her the story of the brave and brilliant Queen Esther, who, with skill, savvy, and intelligence, saved the Jewish people from destruction at the hands of the king's evil and conniving minister, Haman.

Queen Esther was a true heroine and role model. After she told the story, my friend's mother would dress her up as Queen Esther, with a crown and robe and scepter, and

Vicki would spend the day basking in the glow of her royal identity. "My mother made me feel as if I *was* Queen Esther," she said. "Brave and strong. Capable of greatness. And I never stopped believing it."

The foundation of self-esteem is assembled from the building blocks of childhood, with the strongest influence being the parents. Even though some children are able to survive and thrive despite third-rate parenting—perhaps because of some indefinable genetic quality—why should anyone take the chance? In the debate over whether nature or nurture makes a person who she is, I vote for nurture.

SUCCESS BUILDS CHARACTER

Failure doesn't build character. Success builds character. Whoever said, "If at first you don't succeed, try, try again," warped the minds of several generations of parents. People think that their children learn important lessons from failure. I believe a child can learn more in a moment of success than can ever be learned in a month of failures.

One of a parent's many responsibilities is to see to it that a child is given opportunities to succeed in a number of youthful endeavors—learning how to play with other children; learning how to share, read, count, color, and write. So much is going on for children at such an intensive rate. We should all try to restore some of our memories of our own childhoods as a guide for raising our children. Do you remember what you were good at, and what you weren't good at? Did you eventually acquire some of the skills that, as a child, seemed completely beyond you? Make a list of what your child is good at, and nurture those talents. Show love, compassion, and tolerance for your child's weaknesses and failings. Build your child up. Nothing is as searing as a parent's disapproval, anger, or disgust. Call your child an idiot long enough, and what do you think your child will believe about herself?

HIGH HOPES

Children are born with a natural instinct to please their parents. They crave our attention and our approval. Infants are thrilled when we delight at their antics. Little tots go to nursery school and bring home their artistic efforts to be taped to the refrigerator. They make little jewelry boxes for Mother's Day. They are proud when we praise them.

Good parents know how to let their children thrive in their own unique way. I emphasize the word "unique." Our children are not carbon copies of us; nor are they put on earth to fulfill the dreams that we did not attain ourselves. If we let it be known that we expect them to excel in areas that do not interest them, or to follow a path for which they don't have natural ability, they are going to struggle and feel inept. If we voice disapproval, or suggest that they are displeasing or disappointing us, their self-esteem and confidence can be shattered.

Woe to the doctor's child who gets a C in biology, or the mayor's child who fails civics, never mind the tens of thousands of athletes' kids who can't throw a ball or wield a tennis racket.

Make an honest examination of your expectations and your motivations. Take care that your high hopes are not the source of your child's diminished spirit.

THE MODERN EXCUSE

Have you noticed a trend during the past few years? Suddenly, there seems to be an elaborate cabal of physicians, child psychologists, child psychiatrists, and psychotherapeutic counselors declaring that kids are increasingly afflicted with chemical imbalances or learning disabilities. And guess what? They happen to have just the medication for the condition.

Kids today don't have poor study habits. They're "learning disabled."

They don't have discipline problems. They have "ADD." That is, attention deficit disorder.

They're not high-spirited. They're "hyperactive."

They're not having a lousy day. They're "depressed."

The answer? Drugs. Ritalin, Prozac, Zoloft. There's a pill for everything. Can it be that the greatest country in the world has suddenly spawned a whole generation of psychologically and chemically defective children? Clearly, some children require medical intervention. But, by and large, troubled children are usually the products of parents who either wittingly or unwittingly fail to see the signs of trouble early on. Those parents can range from the truly dangerous and psychotic to the well-meaning, well-educated people who deal with any sign of unrealized expectations for their children by blaming their child's brain chemistry. God forbid one's child should just be "average."

I had a case in family court some years ago that I've never forgotten. You might say it's a case in point.

A twelve-year-old boy was brought before me on an arson charge. He and another boy had been arrested for stuffing a discarded mattress into an elevator in the building where they lived, dousing it with lighter fluid, and setting it on fire. They pushed all the elevator buttons, so the elevator with the burning mattress would open at every floor. It caused a lot of panic, smoke, and damage. When I had the boy in court, I looked over his case file. A single mother. Alcoholic. The boy on medications. I began to flip through the years of social services history. The mother had been barely a child herself when she'd given birth. She was drinking, on welfare, bringing the baby to the local clinic for care. By the time the child was two years old, he was an active toddler, curious and alert. There was a note in the file from a doctor the mother consulted at that point. She told him the child was unmanageable; he was everywhere, always up, always moving, always talking. She couldn't handle him. The doctor wrote: "Mother reports hyperactivity, sleeplessness, excitability." An average two-year-old, I thought, bright-eyed and bushy-tailed. But the doctor diagnosed the child as hyperactive and pre-scribed medication to calm him. Medication for the two-year-old son of a single, alcoholic mother who was too wasted to take care of herself, much less her child.

Here's the kicker. The twelve-year-old boy appearing before me on an arson charge was still on medications. He'd been on medications since he was two years old, with negligible follow-up.

I suspended the hearing, moved it to a later date, and had the young boy put through a complete series of examinations. He was slowly weaned off his medications. He blossomed. He became a different child entirely. It was as though a veil had been lifted, a fog cleared away.

I would have liked to keep him in the state's care, but he wanted to go home. "Mom needs me," he said. A good boy. A tragic situation. How many children are medicated because their parents can't deal with them? I have seen parents stand before me in court and swear their children are uncontrollable, unsalvageable monsters, without a normal sense of right and wrong. They swear that their children's judgment and learning abilities are entirely skewed. They demand their children be counseled and medicated to solve their problems. Sometimes medication is the answer, but it's no replacement for an orderly environment in the home and a solid foundation based on positive values.

What of those children who are the victims of unrealistic expectations by parents? Some children are going to be slow learners, awkward, difficult, temperamental, or worse. Their parents, unable to face the truth, look for a scapegoat, and a medically treatable chemical imbalance fits the bill. On the one hand, parents think their children are the second coming. On the other hand, they believe they can give a child medication without the child getting the message, "Something is wrong with me. I'm not normal. I can't do it."

If a child acts out, misbehaves, is not paying attention, runs wild, is a bed wetter, is undisciplined, refuses to cooperate, says horrible things, and uses horrible language,

something may be going on that makes otherwise inex-
plicable behaviors entirely clear. Perhaps the parents are
constantly bickering and fighting. Perhaps there's domestic
violence, physical or sexual abuse. There may be a new
boyfriend living with the mother. Maybe a new baby has
been introduced into the family, or there's been a divorce.
So many factors can cause a child to act out. Explore the
possibilities. Talk, listen, and then listen some more. Be a
parent. Focus, observe, communicate. I'm not saying it's
easy. Parenting is the hardest job in the world. But it is the
obligation of parents to help their children thrive in the
world, and grow into adults who thrive. Let's not start
them off with a handicap. Medicate a child as a last resort.

Bring Your Children to Work

I'm big on bringing children to the workplace. One of the
best ideas to be instituted in recent years is the "Take Your
Daughters to Work" day. There's only one problem with
the idea. It should be "Take Your *Children* to Work."

Why? Because when mothers bring their sons to work,
it can engender in those boys a sense of respect for
women that they'll retain for the rest of their lives. When I
was the supervising judge in Manhattan, I said, "I think it's
outrageous just to bring your daughters to work. Let's

bring all our children to work, so they can see what it looks like to have a work ethic—to be passionate about what they do." When you bring your children to work, it also provides them an opportunity to see if what you do interests them, because if you really love your job, the chances are you'll pass that enthusiasm on to your children, and it will become a skill that they'll be eager to develop. How many physicians, lawyers, and actors have children following the same path? Three of our children have followed us into the law. They saw how much we loved our jobs, and their constant exposure to it revealed an innate ability of their own.

The workplace also exposes kids to the real world, not what they see on a television or computer screen. It's one of the best training grounds there is.

KIDS AS KICKBALLS

Nothing burns me up more than parents who use their children as leverage in divorce proceedings. In my opinion, some of these custody and child support cases amount to child abuse. When I was in family court, I saw men and women resort to disgusting tactics that they claimed were "for the sake of the children." Bull.

You tell me. Is it good for a child to be asked to spy and

lie? Does it make a child feel secure to see her mother bringing false charges of sexual abuse against her father? Is a child's self-worth enhanced when he's used as barter in an adult battle? Is a child's confidence strengthened when she hears one parent call the other unfit?

When I was a family court judge, I was naturally suspicious of allegations of sexual abuse made during bitter divorce proceedings. Occasionally, an accusation of this type would stand up to scrutiny, but more often women were trying to punish their ex-spouses. I was so outraged when I uncovered false allegations that I sometimes ended up giving the fathers sole custody.

Divorce itself is one of the most confusing and terrifying events in a child's life. It takes a great deal of love and support to restore a sense of safety and normalcy. Add an explosive setting, with parents pitted against each other, and a child can be seriously damaged.

When Jerry and I married, we had both been divorced, and there were five children in the mix. Whatever differences might have existed between Jerry and his ex-wife, and me and my ex-husband, we hoped to keep the children out of the fray. We made some mistakes, but we tried. (Believe me, it's the *mistakes* they remember as adults.)

I also believe that for the most part children would be better served if parents, once divorced, could be equal partners in their upbringing. Most of the children of divorce live with their mothers, and the fathers become uneasy intruders or weekend visitors. Children are robbed of the opportunity to develop a really strong bond with a paternal figure.

As adults we have the ability to set aside our grievances and realize that even when we are no longer spouses, we are still and will always be parents. The question is, do we have the *will*?

SEQUINS DON'T BUILD CONFIDENCE

I disapprove of little-girl beauty pageants. I don't believe for a minute that they're about building poise and confidence. What they say to the little girls who are in pageants is that the girls have value because they're pretty. What they say to the little girls who *aren't* in pageants is that nobody will make a fuss over them because they're not pretty. Why do we keep reinforcing the very messages that have made so many of us so miserable? Pageants for boys have to do with skills and sports—being a chess champion, building a science project, hitting a home run out of the park. There are no little boys in beauty pageants.

The tragedy of JonBenét Ramsey brought the message home to millions of Americans when we turned on our TVs to see this little six-year-old girl, in false eyelashes, mascara, and red lipstick, strutting around in feathers and boas, vamping for the cameras. There was a sadness to the scene. No six-year-old would want to do that on her own. Little girls like to play dress-up, but without pressure from an adult, it wouldn't occur to them to be so sensuous. Jon-Benét is an extreme example of what many mothers do when they tell their daughters to fix themselves up and look pretty so they will attract a boy. I don't know any mothers of boys who say, "Go work out at the gym, lose some weight, cut your hair—so you'll be attractive to girls." All boys have to do to be attractive to girls is change their socks periodically.

You can't protect your daughter completely from exposure to the pervasive cultural messages and peer pressure to look and act a certain way, but you can provide strong and consistent alternative messages. Begin by emphasizing abilities, skills, and personal qualities that aren't related to looks. And when you do comment on your daughter's appearance, try to avoid the emphasis on looks as an ideal apart from everything else. It's one thing to say, "Sarah, you've got the prettiest eyes," and another entirely to say, "Sarah, your eyes are so bright and expressive." Subtle distinctions build confidence and place the emphasis on qualities rather than appearances.

It seems unfair that a truly homely man can become attractive based on his professional success or inherited wealth. How he looks doesn't empower him. What he has, and who he is, empowers him.

That should be our goal as women. The truth is that the stunning woman will get some fleeting attention when she walks into a party. It doesn't matter if she's as dumb as dirt. But it's always the interesting, witty, socially adept women who seem to keep most men's attention over the long haul—at least the men worth having.

R–E–S–P–E–C–T

Do you remember the huge Aretha Franklin hit "Respect"? In a lot of ways, that song says it all about the problem we continue to have between men and women, young and old. The Queen of Soul spelled it out: R–E–S–P–E–C–T. As in, treat me with respect. Act respectfully in my presence. Respect my feelings, my wishes, my needs. It's a shame they don't play the song more often—it's a message that still needs to be heard.

Instead, what musical messages are young people hearing today? The lyrics of many songs are filled with violence, profanity, and disrespect for women.

Don't tell me it's "only music." Don't tell me it doesn't make a difference. These words have the power to imprint subliminal messages in unformed minds. Wouldn't you rather those images come from respect?

What's So Bad About a Little Discipline?

You know why there's so much stress on college campuses? Because students don't do their work when they're supposed to. It's as though they can't figure out that each semester is only about three and a half months. By the time they turn around and begin to organize themselves for an all-out, full-scale frontal attack on their studies, they're at the end of the semester with a paper due. And the colleges reinforce the lack of preparation by making it so easy for the students to put off their work. The professors offer extensions on the papers. My niece is just now completing a paper that was due nine months ago. She hasn't turned it in, and she already got a grade on it. Now that's a trick I wish I'd known about when I was going to school!

Lack of structure does a disservice to children. By the time they reach college, it's too late to impart discipline. Structure should start from day one. It helps people cope with their environments. Those people whose lives are all over the place are usually the people who are most stressed. Why? Because life has a way of throwing you curve balls—unexpected crises. And if everything in life that you *can* organize *is* organized—your house is relatively clean, the papers are all where they're supposed to be, you have the birthdays and anniversaries written down, you get up to go to work at a certain time, you get there and park your car in a certain spot, you go in and order the same kind of bagel and coffee in the morning—then there's a certain structure that you live by. It's a safety net. Structure and routine are essential for children. I'm not talking about rigidity or lack of spontaneity. I'm talking about the everyday experience of having life feel secure.

The truth is, kids thrive on structure. It makes them feel safe, loved, and cared for. Yet we ignore this in favor of open classrooms in schools, where dress codes are nonexistent and children flit aimlessly from one activity to another. No wonder we've become a number one country with a second-rate school system. It does not build self-esteem in children when they don't know the boundaries. Be a parent first, a friend second. When my children were growing up, I was not their friend. I had friends of my own, and so did they. Our relationship was simple: I was the adult and I made the rules; they were the children and they followed the rules. There was none of the wishy-washy ambiguity we see so much of today. My son Adam tells a story about when he was five years old and stole a box of cap pistol caps from the store. According to Adam, I dragged him into the store and told the manager, "If you ever see this boy in here again, call the police and have him arrested."

Tough love. Today, Adam is a successful lawyer and a delightful, well-adjusted adult.

PLAY-AND-PLAN

When Jerry and I got married, I had two little children, and Jerry had three. We became an instant big family. Five kids, two parents. It immediately became apparent to me that Jerry and I were to play very different roles in the family structure. Jerry was the yes-man and I was the no-ma'am. At first, I didn't think this was such a great thing. Maybe I was a little bit jealous of Jerry's popularity. The kids loved him—what wasn't to love? He said yes to everything. He was their playmate and their pal. He also provided them with a number of very positive character traits to emulate. Jerry was warm, loving, generous, and fun. I heard the stories about Jerry's mythic status among children even before we were married. When his kids were small, he'd go out early on Sunday mornings to get the papers and some bagels, and he'd always come back with a couple dozen little treats—gumballs, penny candies, tiny wax bottles with syrup inside, whatever—and all the kids in the neighborhood would come around every Sunday morning for a treat from Jerry. It's the kind of thing kids never forget. Meanwhile, when Jerry and the kids were busy wrestling in the mud, playing ball, bike riding, and roughhousing, I was left with the job of being the serious one who took care of the less popular but necessary tasks of making sure the children had regular medical checkups, balanced meals, clothes on their backs, and discipline. Their faces didn't exactly light up when they saw me coming at them with a comb and toothbrush, but somebody had to do the dirty work.

I had always viewed parenting as serious business—and it is—but Jerry taught me that children thrive when they have both a firm hand and an element of playfulness in their lives. Our children, I realized, were very fortunate because they had parents who could give them both. I eventually was able to make peace with the way things were, because it benefited the children. I think both parents have to be honest with themselves and target the things each of them is best at in regard to the kids. It's an exchange of strengths. To be sure, my role wasn't as much fun. Nobody wants to pick up after five kids, much less keep straight their various schedules, and be sure they did their chores, but all that has to be done as well. So I always ended up taking care of the more practical stuff, while Jerry was able to be both father and pal, coming home with surprises and toys and games. I had to hand it to him—he was always consistent, always giving hugs and kisses, always telling the kids he loved them, always showing his feelings of adoration openly to them. He was a real self-esteem booster for our kids. He was the Pied Piper to his children, and he's now the Pied Piper to his grandchildren. When our grandkids come to visit, I'm always left kneeling with my empty arms outstretched, my lips puckered up for a kiss, as they run right past me and into the arms of Grandpa Jerry, his pockets stuffed with bubble gum and lollipops, none of which they're allowed by their parents, of course. That's the kind of grandfather he is, though. He'll buy his grandchildren things that their parents have strictly forbidden. He'll sneak in all kinds of little play toys. For example, little boys reach an age when they're all desperate to get their hands on something that looks like a gun—just the kind of things parents in this enlightened and politically correct age don't want their children to be anywhere near. Jerry will go out and come home with water guns, light sabers, battery-powered facsimiles of burp guns that make the kinds of sounds that thrill little boys—anything to amuse and fascinate his grandkids. I always have to confiscate this cache of toy weaponry before the kids leave our house, lest one of their parents (now in the role of serious adult) discovers the source of this illegally obtained delight. I keep the toy weapons in a closet for their next visit. I *do* get it. I've come to terms with the way it is. Jerry will always provide the exciting stuff—video games, in-line skates, skateboards, water guns. I'll provide all the things that require organization, forethought, and planning. If any of the grandchildren ever want to go on a trip or to a camp, or want a speaker for a school function, they come to me. They want to eat, they come to me. So Poppy provides the fun stuff, Nanna comes up with food, trips, shopping, and anything that requires scheduling. Poppy plays, Nanna plans, and the children get everything they need—just as our children did.

TELL THEM YOU LOVE THEM

I always knew that my mother and father loved me. They didn't say it a lot, but I just knew. And when I had children, I hoped that they knew my love for them was total and unconditional—even though I didn't say it much, either.

Then I met Jerry Sheindlin, and he impressed me. One of the first things I noticed was that he was constantly saying "I love you" to his kids. If he was on the phone with them or kissing them good night, he'd always add "I love you." I began to realize how really nice that was. Even though your children may know that you love them, it's a comfort when you reinforce it verbally.

Saying "I love you" is one of the best habits I picked up from my husband. When I speak to my children today, even though they're now adults, I tell them I love them.

And now I pass that advice on to parents: Take the time when you say good night to your children, when you tuck them in, when they're going off to sleep, or when they're going out on a date, to add "Have a good time. I love you. Speak to you next week. I love you. See you in the morning. I love you." They know it, but it's still nice to hear it. The power of love isn't diminished by the repetition of the words.

CHAPTER 8

Letting Go Is Half the Fun

When I first got married at age twenty, my new husband and I lived in a small apartment in Brooklyn. We were very busy. I was finishing law school, and Ron was working for Legal Aid. There wasn't a whole lot of time to look after all the little housekeeping details, and we didn't have very much money.

My mother, who was an extremely warm and generous woman, didn't want us to suffer. She loved doing for us. Sometimes I'd come home after a long day at law school, and I'd open the refrigerator to find it freshly stocked with food. The freezer would be filled with individually wrapped steaks and chicken breasts, casseroles, ice cream, and other goodies. My dirty laundry would be gone, replaced with neatly folded piles of clothing, sheets, and towels. I'd walk into the bathroom, and a dozen pair of freshly washed nylons would be hanging on the shower rod, drying over the tub. Wonderful, huh?

To be honest, I was ambivalent, and I even started to resent it a little. Didn't my mother think I was capable of handling all this myself? Didn't she think I could manage my own little place? And by extension, my marriage? Did she think she had to buy us food?

One night, after I came home to a sparkling apartment and a fully stocked freezer, I had a nightmare. My mother was washing the bath mat in my bathroom, and I was standing in the doorway screaming, "I can wash my own bath mat!"

It's human nature to resent being given too much. If you grow to rely on a person too much, there is an imbalance of power, and resentment builds in the heart of the recipient. It's important to give children a sense of security as they grow up. It's also wonderful, if you are able, to provide a safety net to your adult children when they are just starting out. Also recognize that it's a wonderful feeling for young adults to make it on their own. Self-assurance is ultimately achieved when they can spread their wings and fly solo. So give with a warm hand and a generous heart, but also know when to let go.

Button It Up

When you get to that age in your life when your children have husbands or wives, take this word of advice from me—button it up! If you see things your children and your son-in-law or daughter-in-law are doing that you don't approve of, keep it to yourself! Believe me, I know the temptation. I have four in-law children, and I see them do things that I would do differently. That doesn't necessarily mean I'm right, but from my perspective, of course, I'm older and wiser. So what? It doesn't matter. Unless you're asked for an opinion, don't offer one. You may think, "If they'd only listen to me, I could save them a lot of heartache." Resist the impulse. Don't open your mouth, raise an eyebrow, or slam down a plate. If you have until now managed to maintain a relationship with your grown children, don't add undue stress to their lives by touching off arguments and loyalty wars through interference. Just remember: Your in-law children are going to make your children pay for whatever you say or do to them. If you get into it with one of your children's spouses over any of a myriad of subjects—child raising, money, personal habits, cooking, cleaning, work, you name it, you're trolling in dangerous waters, and your adult child may be washed overboard in the turbulence.

PRACTICE IN FRONT OF A MIRROR

If the urge to meddle in your adult children's lives is so overwhelming that you just can't help yourself, you may need to practice achieving just the right air of noncommittal, nonjudgmental poise. Stand in front of a mirror and imagine your adult child making a statement that would normally cause you to launch into a tirade—such as, "Guess what, Mom? I've decided to leave medical school and become a stand-up comedian." Keep your face motionless, your eyes blank, your forehead still, and say evenly, "That's nice, dear."

Perform this exercise over and over until you have it mastered, varying the themes:

"I'm becoming a Buddhist." *How interesting.*

"I'm quitting my job to join the foreign legion." *Good.*

"Bill and I have decided not to have children."
Mmmmm . . .

"I'm having a sex change operation." *Whatever, dear.*

In time you'll have trained yourself so well that your adult children will be begging you for a reaction.

THE *MISHPUKAH*

[Mish-puhk-ah, Yiddish. Relatives, esp. by marriage; in-laws.]

When I was a young newlywed, I was close to my family, and Ron was close to his. My mother-in-law and her best friend alternated holidays at each other's homes. Each Thanksgiving, one or the other would have both families to her place. The next year, they'd turn it around. They'd been doing this for years. It was nearing my first Thanksgiving as a new member of their family, and my mother-in-law called me up. "You know, Judy," she began. "Thanksgiving is coming up. You're spending it with *our* family, aren't you? We'd be so terribly hurt if you didn't. This is a tradition with us, you know. This year it's my best friend's turn. What a spread she puts out! So. What do you say?"

Exactly! What could I say? "When do I get to spend a Thanksgiving with *my* family?" When I told my father about my conversation with my mother-in-law, his response was a classic. "They're not rich enough to have traditions. If you're a Rockefeller, a Kennedy, maybe you have traditions. The rest of us, we have get-togethers."

Now that I have married children, the same thing comes up all the time. The scheduling, the excuses, the carrying on, the dashed hopes, the crushed dreams, the bruised egos—oh boy! The children have us, their spouses have their families. Oh, and add this little fillip to the general "where are we going to spend the holidays" mayhem—my two children have their father, and my three stepchildren have their mother. There are all kinds of families, and all different levels and interchanges of families. It makes the Los Angeles freeway seem like easy traveling. Your choices are plain. You can fight it, and make your lives and the lives of your adult children miserable by demanding that they spend their holidays with you. Or you can say, "We'd love it if you were able to spend the holiday with us. Let us know if you're going to be able to make it." The end. Never make them feel as if they must be with you. That really doesn't work the way you want it to. You end up with angry, surly children who feel as if they really can't communicate with you anymore. You also, of course, add stress to their marriages.

I learned this the hard way, through bitter experience. I didn't do the right thing a couple of times. I blew it. I nagged, I demanded, I cajoled, to get my kids here for certain holidays. I saw the results. I got body counts on the aftermath. I learned to back off. Even today, if one of my children says, "I think we're going to spend New Year's Day with Pop," I might slip up and ask, "Didn't you spend last New Year's Day with your pop?" But still, my children know that I'll no longer bust their chops about holidays with Jerry and me, because we all get to see one another a lot. We're lucky, I guess. We like to get together with our children, and the feeling is apparently mutual.

You want drama in your life? You can cry; you can lose sleep over these events, if you really want to. You can carry on: "Where did I fail? My children don't want to be with me, their own mother? They'd rather go to their other family? We're not good enough for them?"

Or you can look at it like this. You're not going to have to clean up, entertain, shop, cook, serve, or do anything else in some endless holiday cycle, either. You're basically free. You know that your children love you. You'll see them soon. So relax and enjoy.

CHOICES

When your children are young, they're with you every day because they have to be. Then, suddenly, they reach the age when they have a choice in the matter, and you wonder, what if they don't choose to be around you? There is a great temptation to campaign for their approval—to win them over with indulgences.

I'm happy to report that our five children have grown up to be interesting, exciting adults. Some are married, some have children, and we're all very close. We enjoy one another. When we got the house by the lake, I made it clear to the children that they were always welcome. The idea was to have a place for the entire family to be together, and Jerry and I were delighted when they took us up on it.

The first six months we had the new place, it was a constant party—barbecues, swimming, tennis, boating. At least for *them* it was a party. I was shopping for groceries, refilling ice trays, doing load after load of wet towels. I began to feel annoyed. If I started the day with a freezerful of ice, I expected the ice cube trays to be refilled as they were emptied. Laundry also became a problem—not just the towels, but the sheets and pillowcases. Our basic family had grown to fourteen. That's a lot of loads. When the weekends were over, I'd go from room to room, stripping the beds, collecting stray towels and forgotten pieces of clothing here and there. I'd get everything into the laundry room, do the necessary number of loads, dry and fold, then bring everything back upstairs, remake all the beds, then resupply the linen closet and bathrooms.

After one whirlwind weekend, as I was hauling towels and sheets to the laundry room at midnight, I said to myself, "Judy, what are you doing?" It was a good question. By my count, there had been ten adults present in the house that weekend. Why was *I* doing all the work? I realized that a part of me wanted everyone to be relaxed and happy. I didn't want to introduce the element of chores. The result was that I was exhausted, overworked, and resentful. This wouldn't do. I decided to level the playing field. The next time we had all the kids up for the weekend, I proposed a few changes. I said, "Listen, kids. When you get up in the morning on the day you're intending to leave, all of you have to strip your beds, throw your stuff in the washer and dryer, and remake the beds. Please do the towels you've used, too." Everyone nodded. No one looked shocked or affronted. I went on. "This is a second home for all of us, so if you want to use the place when we're not here, please feel free to do so. But when you leave, leave it like you found it. Everything neat, vacuumed, garbage properly disposed of, laundry and towels done." Again, there were no looks of shock or bewilderment, or any signs of annoyance. They all said, "Sure."

Did I feel guilty for deciding to lay down the law? A little. I wanted my children and grandchildren always to have a wonderful time when they visited. The problem was that, by allowing them to have pleasure without any responsibility, I had done myself and them a disservice. I felt a lot better after explaining the problem to them than I would have had I repressed all of it. That would have eventually led to some sort of unnecessarily explosive outburst after I'd been pushed beyond the limits of my tolerance. This was a much better way of handling it. Our children still come to the lake. Asking them to do their part hasn't alienated them in the least.

LETTING GO

For so many women, it is very painful to let go of playing a primary role in their children's lives. It is particularly wrenching for women who have come to judge their own self-worth through their children. When the kids fly the coop and leave an empty nest, they're suddenly confronted with themselves, and they find themselves wanting. They fear that now no one will need them, that they no longer have anything to offer. That's the furthest thing from the truth, of course.

If you're in this position, you may need to spend some time getting to know yourself again. You were a person before your children were born, and you're a person now. Who is that person? How do you find her?

You find her in work. You find her in engagement. That might be a job, a profession, volunteer work, community service, travel, education—whatever fulfills you. You can be engaged in the world at any age. My eighty-six-year-old aunt Lillian volunteers in a senior citizen's home three days each week. She is still as bright as a new penny and as sharp as a tack, and her children adore her. They also admire her.

The truth is, my children resented my career when they were younger. I wasn't always available for them, as were so many of the other mothers, and I admit that I had some guilty moments. It's inevitable in the life of any hardworking career mother, I suppose, but it really wrenched at my heart on more than a few occasions. However, in what seemed like a blink of an eye they became teenagers, and entered that phase of their lives when they preferred that their parents be neither seen nor heard.

As adults they are so proud of their mother, and they

enjoy my newfound celebrity as well. There are a lot of fun perks attached to being Judge Judy's kid now. I believe that they love me, and I know that they respect me. We've been through a lot together. In the process, my daughters and, I hope, my granddaughters have seen that women can play many different roles. However, at the end of the day, no matter how many roles you've had to play, you're always *you*.

CHAPTER 9

You Can Be
the Hero
of Your Own Story

Unless you believe in reincarnation, this is your one shot. (Even if you believe in reincarnation, you could come back as a squirrel or as a lemming, so this is *still* your one shot.) Men seem to understand this much better than women. How many men do you know who put their lives on hold because they're waiting for Ms. Right? I'll bet none. That doesn't mean that men don't want to be part of a couple. It's just that they don't believe in hanging out in limbo until that happens.

Women, on the other hand, because they often feel that their lives are not complete until they are mated, will put off making themselves content in the meantime. I have known women in their thirties who will not buy a couch, let alone a home, because *he* (whoever *he* may be) might not like the pattern. I have also known women in their sixties, divorced or widowed, who won't paint the ceiling, replace the carpet that the dog has chewed to shreds, or move into a bigger, better apartment until they're sure they won't be getting remarried. All these women, young and old, say the same thing: "Well, I'm probably going to meet someone and move in with him, anyway." Even if they don't say it out loud, they say it to themselves, and there is something very sad about that.

Sure, you can hope for something else. If you really want to be part of a couple, there's nothing wrong with hoping for that. But whoever said that you have to live in sackcloth and ashes while you hope? Why can't you have hope while you're living wonderfully? Hope can occur in a nice apartment.

This is a hard message to get across, because even though most young women don't actually *have* hope chests anymore, they have retained a place in female mythology. The hope chest, you'll remember, was the receptacle for the most beautiful linens, the most lovingly crafted quilts, the silver goblets, the heirlooms. All locked away until that glorious day when the hope was fulfilled in the form of a man. Today's variation on the hope chest is the bank account you never touch to buy a new set of towels, even though the old Snoopy set has been with you since you were eight. It's the crooning voice in your head—"*You're nobody till somebody loves you.*" And meanwhile, the clock is running on your one precious life.

So if you're sitting around waiting for the hero who will complete you in every way to show up, maybe you should consider this radical idea: Why don't you decide to be the hero of your own story? Instead of looking for rescue from the outside, look on the inside. Instead of waiting for Prince Charming to sweep you off your feet, you do the "sweeping" by dazzling yourself with your achievements. Instead of just getting by in the hope that something better will come along, live your life as though the better is already here.

GET A LIFE

If you are going to live heroically—that is, to your full potential—you can't afford to waste energy on pettiness. One of the most common types of cases that comes to

Judge Judy's court involves women who are carrying silly grudges. This is the stuff of molehills-to-mountains. The women may have been best friends for forty years. Suddenly, it doesn't matter, because "She crushed my favorite hat . . . She broke my chair . . . She put a scratch on my car . . . She never paid me back the $10 I lent her." Or the people, once in "love," who will spend a year litigating over who paid the last electric bill. I marvel at the consuming passion that can be devoted to disagreements of the puniest proportions.

I hate to see it. People waste precious years in court, harboring grudges, nursing old wounds, seeking revenge for real or imagined slights. Their lives are passing them by; they never get those years back.

Becoming bitter over *bupkis* (nothing) is absurd.

MERRY WIDOWHOOD

The statistics are clear. Women outlive men. There are far more widows than widowers. And those widows aren't standing around with one foot in the grave. They might have another twenty or thirty years of vitality and adventure.

Which is another way of saying that, even if you have the best marriage God ever made, there is a chance that you'll spend some of your years alone. Have you thought about what you're going to do with all that time? Maybe a week, or at most a month, can be spent crying, "Oh my God, Irving's dead. What will I do?" And then, as they say, life goes on.

Many women, when they become widows, suddenly wake up to find that they're themselves again. And although that doesn't mean they don't grieve for their loss, they learn that it's not such a bad feeling.

If you're lucky enough to have good health and resources, you may be able to do some of the things you never had the opportunity to do before. Maybe you've always dreamed of traveling, but your mate's idea of a vacation was setting up the barbecue and the croquet set in the backyard. Maybe you've sublimated your natural sociability for years because the very suggestion that you might have friends over for dinner sent your mate scurrying for cover in the basement. He would have skipped his own wedding if he could have signed an absentee ballot.

Maybe you were perfectly in sync during your married life, but you weren't born coupled. Two individuals come into a marriage. You pack away some of your own dreams to be part of the team. Now the team is retired, and you can dust off the box, open it up, and see what's in there.

You may need to learn all over again your value as a single person.

THE PITY PARTY

So life has thrown you a big curve, and you just don't know how you can go on. The older you get, the more curves there are—physical, mental, emotional, financial. It may feel as if you have no options. But there are always options—and I like this one.

Throw a Pity Party. This concept was introduced to me by my friend Vicki, who heard about it from another friend. Vicki's buddy had one of those big-league curves thrown at her: She was diagnosed with an inoperable brain tumor. Her only hope, and it was a slight one at best, was an aggressive campaign of radiation therapy. "I was impressed with her spirit," Vicki told me. "She was upbeat, she didn't sit around moping and crying—which is what most people would do. So I asked her how she kept so positive. Here's what she told me. She said, 'Every day, from 3 until 3:15 P.M., I have a Pity Party for myself. No holds barred. I scream, I cry, I throw stuff around, I feel sorry for myself. I get it all out. For fifteen minutes. Then I blow my nose, dry my eyes, reapply my makeup, and get on with my life.'"

I thought the Pity Party was such a wonderful idea that I'm sharing it with you. Do you feel sorry for yourself? Has life thrown you a curve? (It doesn't have to be a brain tumor; a bad day at the office will do.) Do you feel like crying, complaining, going to bed and pulling the covers over your head? Save it for your Pity Party. I've tried it. It works.

THE MAY-DECEMBER SYNDROME

Growing older can be a real challenge to a woman's self-esteem, because in this society we have a "shelf life" that is distinctly shorter than men's. You can let it eat at you, or you can eliminate the aggravation altogether. How? About ten years ago, a number of our old friends who had gone through divorces started remarrying. Almost invariably, our women friends who remarried found partners who were roughly in their own peer groups. The men were a different story. Almost all our men friends found women who could have passed as their daughters, and in some instances were younger than the men's actual daughters! It was like a club. You had all these fifty-five-year-old guys with twenty-six, twenty-seven-year-old wives. Great. Good mix.

They'd call up and say, "Hey, let's get together!"

I was always nice, always polite. I'd ask, "So, how old is the new Mrs. Entwhistle?" They'd respond with the age of some of my clothes. My response was always the same. "I don't know. She sounds pretty young. Call me back when you're with someone who didn't just come from her high school prom."

I know that sounds petty of me, but frankly, I don't care. I don't need the grief of spending three hours sitting across the table from some twenty-five-year-old bombshell without an ounce of cellulite, struggling to find a morsel of common interest. I already know we have nothing in common—except perhaps our plumbing. I think the reason many older men are looking for younger women is simple—they want to be in control. They want to be the authority, the one in charge, the general in command. They want a woman who will look up to them, who doesn't so easily see through the facade. And as women get older, they become more powerful, more lucid, more sure of what they want. And that can be very challenging for a man. Too challenging perhaps.

And what's in it for the younger woman? Some young women have told me that they find young men clumsy, brutish, and insensitive. An older man, they claim, knows how to treat a woman. An older man has a lot more to offer. (Especially if he's wealthy.)

So there are a number of sides to this story—the older man's viewpoint, the younger woman's viewpoint, and the older woman's viewpoint.

By the way, when I hear women ask, "Where are all the good men?" I have to reply that they're not paying attention. The good men are right there where they've always been. You just keep missing each other while running around trying to find Mr. or Ms. Perfect.

P.S. I have a lovely, mature, unmarried son. Send letters and photos.

AND THANKS TO THE MOVIES . . .

I can *really* get angry going to the movies these days. I go to see *Bulworth*. Male romantic lead, Warren Beatty, sixty-one. Female romantic lead, Halle Berry, thirty-two.

I go to see *As Good as It Gets*. Male romantic lead, Jack Nicholson, sixty-one. Female romantic lead, Helen Hunt, thirty-five.

I go to see *Six Days, Seven Nights*. Male romantic lead, Harrison Ford, fifty-six. Female romantic lead, Anne Heche, twenty-nine. Sean Connery, sixty-eight, is currently filming a movie with a twenty-eight-year-old female romantic lead. Michael Douglas, fifty-four, with a twenty-five-year-old romantic lead. And there's always Woody Allen.

Do you see a trend here? I like Meryl Streep's take on the older man–younger woman trend in movies. Streep is a beautiful, talented forty-nine. She is a brilliant actress, audiences love her, yet she has trouble finding parts. At a recent awards luncheon for Women in Film, Streep called on filmmakers to stop feeding the "myth that it's a good fantasy for a girl to want to grow up, stop eating, and at twenty-five marry a sixty-year-old and have a fabulous ten years escorting him into his dotage. That's a time-honored fantasy for *him*. What's *hers*?"

Come on, guys, be honest. Older, *poor* men never land the lushest twenty-five-year-olds. You're much taller and more attractive when you're standing on your wallet.

AGING GRACEFULLY

So far I've been lucky. My husband and I are still relatively young, and neither of us has any major health problems. Maybe we'll live to be one hundred! But I've seen my parents age and pass on. It's tough. Perspectives change as you grow older. The things that were of vital importance to you at one point in your life seem to fade into oblivion. And there are endless surprises. There's really almost never any way of gauging exactly what the future will bring. That sense of excitement and adventure is palpable in my life, certainly. If you'd told me five years ago that I'd be working as a judge on television, I would have told you you were crazy. I would have been wrong. There are plenty of thrills left in my life. Maturity has many advantages, none of which are particularly extolled in our youth-crazed society, but I have hope that we can teach these youngsters a lesson about how much better and more satisfying it can be when you've got a little living under your belt.

STANDING TALL

When I'm in New York, Jerry and I often have dinner at a pizza joint across the street from our apartment. (Did I mention that I no longer cook?) It's quick, it's easy, it's the way we like it. One night, Jerry and I were in the pizza joint grabbing a bite, and I got up and went over to the soda case, took out a drink, and started walking back to our table. A well-dressed man in his late fifties, early sixties, was sitting nearby. As I passed his table, he said, very seriously, "I watch you on television. You're pretty good." Then he delivered the inevitable punch line: "But I thought you were much taller." Perhaps he thought he was paying me a compliment. Do you know how many times I've heard this? Fine. So I'm tiny. I'm marginally north of five feet, but I have presence—especially on the bench. It adds height. I looked at this guy chomping on his pizza, clearly expecting me to answer for my newly shrunken state. So I said, "I am tall. But I give a shorter appearance in person." Then I smiled and walked away. The guy looked as if he'd been hit with a curve ball, which he had. Jerry was cackling loudly at our table. And I drew myself up to my full height, which in spirit, anyway, is at least ten feet tall, and rejoined my husband.

I was born short, but I can stand tall. In the story of my life, I am the hero. I am the one who saves the day—not with brute strength but with wit and intelligence.

Ask yourself what it would take for you to feel that way about yourself. It doesn't matter if you're twenty or sixty, alone or part of a couple. Everyone else in your life is merely a supporting player in your one big starring role.

If you have a healthy spirit, a positive outlook, and a sense of personal accomplishment and importance, you will be a better partner, friend, and mother. It's not about being selfish, it's about being the best you can be and rejecting those who would keep you down. It's about being responsible for your own happiness. It's about living your life free of fear.

CHAPTER 10

A Final Word:
Do Not Abdicate!

Women defer to men for one reason—*fear*. Self-esteem is the ability to conquer that fear, to be able to say, with your head held high, "I am not afraid."

"I am not afraid of being alone."

"I am not afraid I'll make a mistake."

"I am not afraid of being disliked."

"I am not afraid I won't be accepted."

"I am not afraid of being unmarried."

"I am not afraid of growing older."

"I am not afraid of being myself."

There is someone very close to me who is married to a lovely man, except for one thing. When he gets angry about something, he pulls a black cloud over his head, and he broods for days, even weeks. She would prefer it if he would scream and shout. The silence is just deadly.

For many years, she so dreaded displeasing him and setting off a three-week cold war that she allowed him to exercise a form of control over her. If she had a complaint that might lead to a fight, she kept it to herself. If she wanted something that she knew he would disagree with, she decided to do without. It wasn't worth it.

A funny thing happened as my friend reached her middle years. She grew into herself. Her business took off, and suddenly she had money and she had confidence. She used the money to purchase a small country house. She used the confidence to go there whenever her husband started freezing her out. She no longer tiptoed around the house or bore the dreadful tension of his silence.

This is what she told her husband: "At this stage of my life, I am not prepared to let you control me. I don't love you any less. I'm just saying I am not afraid. If you're going to freeze me out, I'm not going to sit and suffer. I'm going to the country!"

Think of your own life and the aspects that make your stomach tighten into knots. Be honest with yourself. What is the fear that holds you back? Is it the fear of being alone? Is it the fear of having no money? Is it the fear of growing older?

Those are all general fears. Go deeper and isolate what the real fear is. If you say your fear is of being alone, what is it about being alone that specifically worries you? Is it not having children? Is it boredom? Is it having no one to take care of you in times of need? Is it that others will think you're less important if you aren't part of a couple?

What would it take for you to not have that fear all your life? Be proactive. If you long for a child, there are ways to have a child. If you are bored or need support, develop a close circle of friends. If you worry about what other people think, work on liking yourself instead of trying to please others by being something you're not. If you're afraid to speak up at work and give full voice to your opinions and ideas because you think you'll look silly, do your homework and practice in advance so you'll speak articulately and with confidence.

It's the human condition to get butterflies in your stomach before a big moment, or to feel an instant of dread before you make an important decision. What distinguishes a successful human being from one who is not successful is the ability to work past the fear.

I may come off as fearless on TV, but there have been many occasions in my life when I've felt that glimmer of fear. I'll tell you, I'm much prouder of the times I walked through it than I am of the times I held back. I'm far more satisfied with the outcome when I've bristled with confidence than when I've abdicated my needs and desires for the sake of being liked.

My husband is an expert on the subject of DNA. He has written two books about it, and he teaches a class on forensic DNA to law students. Jerry can extemporaneously expound on the intricacies of genetic material, with terms that make my eyes glaze over.

Well, I have now written a book on DNA as well—it's this one. Except my book has nothing to do with genetic characteristics, which you can do nothing about, and everything to do with your life, which you can control. In *my* DNA book, the letters stand for:

Do Not Abdicate

"[Judge Judy is]. . .
part Harry Truman, part Rhea Perlman:
funny, quick-tempered, bluntly honest."
—*People*

Listen to

the *New York Times* bestseller

BEAUTY FADES, DUMB IS FOREVER

as read by the author

JUDGE JUDY SHEINDLIN

---

3 hours/2 cassettes
ISBN 0-694-52084-5 • $18.00 ($26.50 Can.)
abridged

Also available in a large print edition from 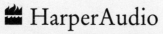 HarperLargePrint

Available at bookstores everywhere, or call 1-800-331-3761 to order.

 HarperAudio
A Division of HarperCollins*Publishers*
www.harperaudio.com